Cloud-Based Music Production

Cloud-Based Music Production: Sampling, Synthesis, and Hip-Hop presents a discussion on cloud-based music-making procedures and the musical competencies required to make hip-hop beats.

By investigating how hip-hop producers make music using cloud-based music production libraries, this book reveals how those services impact music production *en masse*. *Cloud-Based Music Production* takes the reader through the creation of hip-hop beats from start to finish—from selecting samples and synthesizer presets to foundational mixing practices—and includes analysis and discussion of how various samples and synthesizers work together within an arrangement. Through case studies and online audio examples, Shelvock explains how music producers directly modify the sonic characteristics of hip-hop sounds to suit their tastes and elucidates the psychoacoustic and perceptual impact of these aesthetically nuanced music production tasks.

Cloud-Based Music Production will be of interest to musicians, producers, mixers, and engineers and also provides essential supplementary reading for music technology courses.

Matthew T. Shelvock teaches at Western University in Ontario, Canada. Matthew currently works in the music industry in Legal and Business Affairs and also has professional experience as a session musician, engineer, and producer.

I0894746

Perspectives on Music Production Series

Series Editors
Russ Hepworth-Sawyer
Jay Hodgson
Mark Marrington

Titles in the Series

Mixing Music
Edited by Russ Hepworth-Sawyer and Jay Hodgson

Audio Mastering: The Artists
Discussions from Pre-Production to Mastering
Edited by Russ Hepworth-Sawyer and Jay Hodgson

Producing Music
Edited by Russ Hepworth-Sawyer, Jay Hodgson, and Mark Marrington

Innovation in Music
Performance, Production, Technology, and Business
Edited by Russ Hepworth-Sawyer, Jay Hodgson, Justin Paterson, and Rob Toulson

Cloud-Based Music Production
Sampling, Synthesis, and Hip-Hop
Matthew T. Shelvock

Cloud-Based Music Production

Sampling, Synthesis, and Hip-Hop

Matthew T. Shelvock

Routledge
Taylor & Francis Group
NEW YORK AND LONDON

First published 2020
by Routledge
52 Vanderbilt Avenue, New York, NY 10017

and by Routledge
2 Park Square, Milton Park, Abingdon, Oxon, OX14 4RN

Routledge is an imprint of the Taylor & Francis Group, an informa business

© 2020 Taylor & Francis

The right of Matthew T. Shelvock to be identified as author of this work has been asserted by him in accordance with sections 77 and 78 of the Copyright, Designs and Patents Act 1988.

All rights reserved. No part of this book may be reprinted or reproduced or utilised in any form or by any electronic, mechanical, or other means, now known or hereafter invented, including photocopying and recording, or in any information storage or retrieval system, without permission in writing from the publishers.

Trademark notice: Product or corporate names may be trademarks or registered trademarks, and are used only for identification and explanation without intent to infringe.

Library of Congress Cataloging-in-Publication Data
Names: Shelvock, Matthew T., author.
Title: Cloud-based music production : sampling, synthesis, and hip-hop / Matthew T. Shelvock.
Description: New York : Routledge, 2020. | Series: Perspectives on music production | Includes bibliographical references and index.
Identifiers: LCCN 2019045303 (print) | LCCN 2019045304 (ebook) |
ISBN 9780815353195 (paperback) | ISBN 9780815353188 (hardback) |
ISBN 9781351137102 (ebook) | ISBN 9781351137096 (adobe pdf) |
ISBN 9781351137072 (mobi) | ISBN 9781351137089 (epub)
Subjects: LCSH: Rap (Music)—Production and direction. | Cloud computing. | Software synthesizers.
Classification: LCC ML3531 .S46 2020 (print) | LCC ML3531 (ebook) |
DDC 781.64/9134—dc23
LC record available at https://lccn.loc.gov/2019045303
LC ebook record available at https://lccn.loc.gov/2019045304

ISBN: 978-0-8153-5318-8 (hbk)
ISBN: 978-0-8153-5319-5 (pbk)
ISBN: 978-1-351-13710-2 (ebk)

Typeset in Times New Roman
by Apex CoVantage, LLC

Visit the eResources: www.routledge.com/9780815353195

Contents

Introduction

Cloud-Based Music Production (CBMP)

This book covers the topic of cloud-based music production (CBMP) tools as they are used within today's hip-hop beat-making scene. A host of these CBMP services, which support the creation of sample-based and synthesizer-based music, have recently become widespread in the music production community. As a result, this book focuses on key CBMP features which shape the way that music producers work. Some of these features include easily searchable sample libraries, which contain sounds made by professional producers; a host of Digital Audio Workstation (DAW) plugins, such as synthesizer VSTs and mixing plugins (as well as presets for these tools); and, in some cases, these services can offer free recording software (i.e., DAWs), collaboration tools, contests, and other forums for online community engagement. Throughout this book, I explain CBMP music-making methods using both text and audio examples. Readers are encouraged to listen to these audio files, which can be found on this book's webpage on the Routledge website (www.routledge.com/9780815353195).

CBMP services are extremely popular in today's music-making communities, and, at the time of writing, one platform known as Splice boasts over 2.5 million subscribers (King 2019). This means that over 2.5 million music makers are currently *creating* music using only one of the available CBMP platforms. Several other well-known services exist, such as Loopcloud, Noiiz, and 99 sounds. Thus, the CBMP community accounts for a large portion of the current music-making population, and, in fact, CBMP skills have become a new standard competency for many contemporary producers. Although creatives working in numerous genres use CBMP tools, these technologies have been particularly appealing for hip-hop producers who, historically, were known to sift through crates of vinyl records in order to find audio samples from which they could create beats. Contemporary beat makers have since added a number of new techniques to their repertoire because of advances in digital recording tools and cloud-computing. This book clarifies the ways that CBMP tools support the creation of hip-hop beats in more current studio settings, such as modern project studios.

The topic of hip-hop production also has a special relevance at the moment since hip-hop is extremely popular for music fans. One out of every three music streams in the USA in 2018 were from hip-hop records (Resnikoff 2019). In the same year, streaming services amassed a total of 901 billion total streams, meaning that hip-hop streams accounted for approximately 303.3 billion streams (Caulfield 2019). As a result of this recent boom in hip-hop streaming, this book focuses on current (c. 2009–2019) personal computer-based hip-hop music and production techniques which largely support the creation of this new music, rather than on the DJ-centered production styles of the classic hip-hop era (c. 1980–1999). However, in some cases throughout this book where these older techniques are relevant, I explain the influence of classic hip-hop sounds on today's hip-hop production scene since, as readers will learn, vintage hip-hop sounds remain sonically influential as the genre continues to transform.

0.1 DEFINING MUSIC PRODUCTION

For both scholars and musicians, confusion abounds on the topic of music production. And who could blame them? In the past, professional record creation occurred almost exclusively in expensive private studios, and, apart from those directly involved on a given recording project, few were able to witness the actions of legendary music producers such as Phil Spector in the 1960s, for example. In other words, historically speaking, music production was a more private affair, and, as a result, very few people had access to studios and recording equipment.

However, in the 1980s, this changed when recording studios began to incorporate digital technologies, and, by the late 1990s, music production equipment became much more readily available to all musicians. Cumbersome analog processes which required bulky hardware were digitally mimicked—large (and expensive) racks of analog gear were now accessible through a computer screen. Large groups of musicians took note, and, as a result, they began to create their own records. In gaining access to recording tools, home (and project)-based recordists could now easily execute standard record production techniques; for many of these producers, the home studio became a gateway to the professional music industry. As a result of this increased access to equipment, amateurs and semi-professionals were now able to participate in music production *en masse*, and, at the same time, recording skills became a major aspect of one's musical literacy—a rite of passage to the professional music world. Very quickly, machine-based music making became a lingua franca across nearly all commercial music genres.

During this time, the music industry also began to see the of a number of independent music aggregators such as CD Baby and, later on, Soundcloud and Bandcamp. Now, musicians could easily record and release their own music. New production styles, methods, and sounds began to emerge as a consequence of this digital studio boom, which began in the late 1990s and early 2000s. All of a sudden, the term *music producer* could refer to a wunderkind like George Martin—the so-called fifth Beatle—who is famous for his mastery of instrumental arrangement, songwriting, and record production skills, or it could refer to an electronic musician like

DJ Tiesto, whose command of newer musical techniques and technologies earns him a nightly gross of $250,000.[1] Today, musicians like Tiesto primarily create records from their own home studios. While home recording was once exceptional, today it is the norm.

Music production continues to change and evolve, and, as a result, the term is used in more than one way (Hepworth-Sawyer & Golding 2011: 3). However, recent formal research has caused more confusion regarding the term *production*, as many scholars who do not participate in record production define the term to mean "everything done to produce a recording of music." This overly inclusive definition does not withstand analytical scrutiny nor does it resemble (or clarify) how actual music production cultures use the term (Hepworth-Sawyer & Hodgson 2017: xii). It remains unclear how such a broad definition can benefit recordists or researchers who study actual record production activities, especially when the term already refers to a number of clearly identifiable activities.

In order to refocus how scholars use the term music production, I have suggested elsewhere that any reasonable definition of music production must consider how the term is used by music producers and others who participate in the culture of record production because, for these people, a *producer* is someone who demonstrates a specific set of musical competencies pertaining to the *production* of a record (Shelvock 2017a, 2019). These competencies differ from other types of collaborators, such as session musicians or other performers, insofar as a producer is the person who is responsible for overseeing the encoding of a musical experience within a replayable document. In fact, the operative structure of the recording industry, like any other, is wholly dependent on the contributions of skilled workers of various types. Various personnel provide the recording industry with a number of different services from business tasks concerning administration and sales to creative tasks such as production and performance. As researchers, it is important we understand these roles as they exist within actual (i.e., observable, knowable, or otherwise well-documented) music scenes.

At the outset, for some readers, production may seem like a fairly straightforward task: record producers create records. However, to make matters confusing, a record producer's skillset might include the application of a number of traditional musical competencies during the production process, such as songwriting, arranging, or composing music—so long as the producer does these activities for the purpose of creating a record. In order to create records, one has to understand how to convey a musical design—whether we call this design an arrangement, song, composition, or beat—using the medium known as recording. The creation of recordings requires familiarity with processes such as tracking, editing, mixing, and mastering; in hip-hop, sampling and synthesis are also crucial techniques.

Producers may possess any combination of the competencies discussed above, and, in turn, a producer's skillset dictates how one works. In their book *What Is Music Production*, Craig Golding and Russ Hepworth-Sawyer suggest that engineers can be broadly classified into two camps: engineer-producers and musician-producers (2011: 3–9). Musician-producers often hire trusted engineers to handle production tasks which take considerable time and energy, and they tend to focus on songwriting, arrangement, coaching musicians, and personality management. Engineer-producers, on the other hand, tend to focus on recording, mixing, synthesis, sampling, and other studio effects, and these producers may hire outside help for instrumental arranging, orchestrating, or even songwriting. However, both types of producers possess skills as both *musicians* and *engineers*, even though their practices may favour one area over the other (2011: 8–9).[2]

In hip-hop, most (if not all) producers fall into the engineer-producer category, and they make beats primarily using digital recording platforms, sample-triggering devices, pre-recorded audio (i.e., samples), and MIDI. At the same time, numerous episodes of *Rhythm Roulette* demonstrate that producers such as Rahki (Top Dawg Entertainment), Jake One (G-Unit, 50 Cent, Brother Ali, MF DOOM), and Brasstracks (Chance the Rapper) also freely incorporate instruments they play themselves, such as keys, bass, and guitar.

These sounds, whether they are collected via sampling techniques or live recording, are almost always collated within a digital audio workstation (DAW) of some type. Some popular examples for hip-hop production are Ableton, Logic, Pro Tools, and Fruity Loops, although others are used as well. During record production, the DAW becomes a partner to the producer throughout the creative process. Practically speaking, this is accomplished in two ways. When using a DAW, recordists can experiment with endless permutations of sounds, instruments, rhythms, and melodic or harmonic events, and all of these modifications can be compared, or *auditioned*, at any time. In fact, the DAW itself is considered a crucial instrument for songwriting by many, as Mark Marrington writes (2017: 77):

> Like the piano or guitar in previous eras of songwriting, [the DAW] is an instrument in its own right which impacts upon the conception and organization of musical ideas. To put it another way, the DAW has its own particular creative "paradigms" to contribute to the songwriter's process, which once understood, can be harnessed to great effect.

And, indeed, the DAW has had a remarkable impact on songwriting, arrangement, and composition, in addition to its primary functions as a tool for tracking, editing, mixing, and mastering records. This is particularly true in hip-hop, where DAWs are used as a cornerstone for beat-making.

Since hip-hop records accounted for more than 1 in 3 of all music streams in the USA in 2018, it is safe to say that the music-making techniques used to create these tracks are one of today's most relevant musical activities. Yet, as it stands, music scholars and educators know little about this process. When authors do broach issues surrounding hip-hop production, such as sampling, they typically fail to discuss the technique in a way which resembles what musicians and producers actually do in studios.

Instead of directly discussing these practices, most authors instead focus on the application of cultural or sociological theorization to the object or music production. In fact, in most cases, published accounts of sampling and production are entirely disconnected from the practice of beat-making *per se*, as it occurs within contemporary hip-hop music-making subcultures. Notwithstanding this, in rare cases wherein hip-hop beat-making receives a more methodologically rigorous treatment by authors, technological practices (i.e., production activities) are often *black boxed*, or these texts simply do not cover the act of music production *per se*. Quite often, such studies provide an excellent post hoc analysis of records, but they do not set out to directly address the techniques that created these records (Sewel 2013, 2014; Schloss 2014).

Instead, in this book, I focus my analysis on the creative inputs which *create records*, rather than the identification of these same techniques *on available records*, like so many other papers and books have already done. Readers should be aware that simply aurally identifying the use of audio samples on hip-hop records, for instance, has little to do with the analysis of how a sample was (i) captured, (ii) edited, (iii) sonically altered (i.e., mixed), and (iv) repurposed within a new composition or arrangement. As beat makers know, each of these production steps is rife with sophisticated aesthetic nuances, and a record's final sonic characteristics can dramatically change during any one of these stages (i–iv). Until researchers establish an account of these studio techniques in way which actually reflects their usage within music-making communities, we will not be able to substantively or accurately critique them.

An entire subfield is dedicated to the accurate discussion and analysis of recording practice, and more authors are beginning to discuss the practice of music production *per se*. The emergence of music production studies (MPS) in 2017, as it is called, has begun to cause scholars to evaluate the competencies required to create records. This book, of course, is aligned with the goals of this subfield. A key difference in MPS, versus previously available analyses of records and studios, is that music production scholars directly analyze *production*, or the creation of records. Quite often, these scholars possess training, or experience in executing, the music production techniques which they discuss. Since we are only now beginning to see the emergence of accurate music production research, it makes sense that researchers with experience in professional production scenes are currently leading the charge, so to speak. And, since these recordist-researchers actively make (or made) recordings, they are better positioned than other types of scholars to discuss the topic. As the emergence of MPS demonstrates, this type of experiential scholarship is on the rise within music departments.

Later in this book, I demonstrate a variety of beat production techniques and case studies. Researchers and musicians will find these cases useful for understanding how hip-hop beat producers *use* tools and techniques such as synthesis, sampling, MIDI, sample-triggers, sequencers, and signal processing devices. By providing these demonstrations of professional productions, I hope to elucidate a number of under-discussed production practices which are extremely prevalent in today's music scene. To clarify, my goal is not to cause the cultural *uplift* of these techniques in any way. Rather, the goal of this book is to prompt scholars and musicians to begin to engage with a number of musical instantiations that are as common as cadences in classical music or fretboard-tapping and power chords in heavy metal (Hodgson 2011).

0.2 CBMP SERVICES

Cloud-computing is a popular information technology systems paradigm wherein users are given access to remote networks which store and manage data, rather than a local server (or one's own computer). Cloud-based music production (CBMP) services use cloud-computing technologies to provide recordists with access to large libraries of audio samples, synthesizer presets, VSTs (including both synthesizers and mixing tools, in some cases), partially completed works, or completed works for remixing (Figure 0.1). In addition, some CBMP services such as Splice also allow users to automatically back up their work on the host's servers, as well as providing web-based beat-making tools. The most prominent cloud-based music production subscription services are Splice, Noiiz, and Loopcloud.

All of the major CBMP services offer what they refer to as *royalty-free* audio samples, which users can freely incorporate into their songs without fear of copyright litigation. In other words, these samples are free of the typical master/publishing licensing restrictions which accompany commercially available music. This term, which is commonly used in CBMP communities, describes the lack of licensing restrictions in connection with these recordings. So, users can freely incorporate these recordings

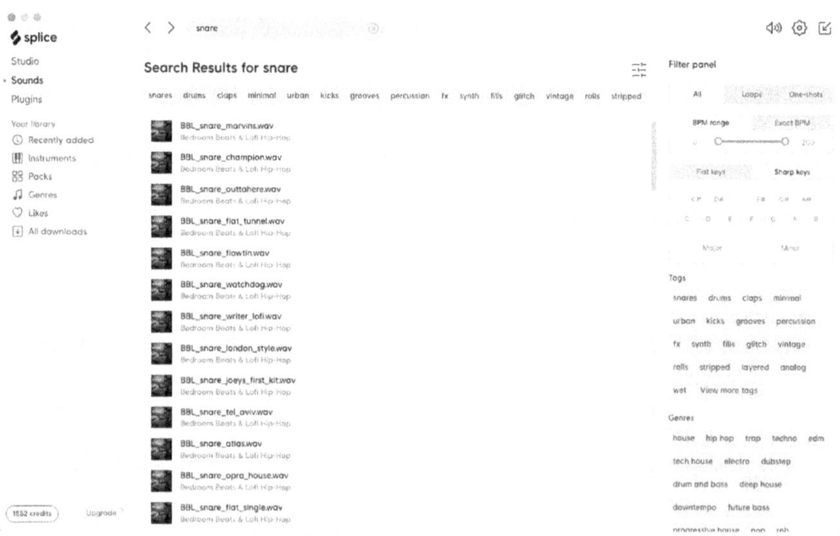

Figure 0.1 Illustrates the GUI in Splice.

(or segments from them) within their work in almost any way they would like to do so.

These *royalty-free* sample libraries receive use from a fairly wide audience. Splice, for example, has over 1 million subscribers. And, at the time of writing, 118 record labels provide Splice with numerous types of audio samples.[3] In addition, users have access to synthesizer presets. Splice co-founder Steve Martocci explains their sample library, stating (in Pangburn 2016):

> When most people think of "samples" they think of stealing a piece of audio from someone else's song and using it without their permission, but it's much wider than that. There are entire labels that fund musicians to produce samples designed to be sold in royalty free "packs". This means as long as you pay for the sample, you can use it in any of your musical work without paying any fees back to the sound designer. We've seen more and more artists making significantly more money selling samples than they make counting on their music sales or streaming revenues. These sounds range from "one shots", which could be a recording of someone hitting a snare drum, to loops which could be 30 seconds of a full drum kit. The vocal samples have also been very popular on the platform—they help give your song an organic feel. Splice Sounds also has presets files for popular virtual synthesizers like Massive and Serum.

Indeed, a popular, yet false, belief persists that sampling audio always constitutes an act of theft or social resistance of some type. While this may have been true in the past, this is not the type of sampling activity that most contemporary (c. 2009–2019) hip-hop producers primarily engage in—although it may happen on occasion. As Noah 40 (Drake, Lil Wayne, Action Bronson) explains, "I'm a fan of publishing, so I try not to sample [from copyrighted material] *too* much, but it's a tool in your arsenal."[4] In addition to Noah 40, numerous other famous producers also incorporate CBMP-based royalty-free samples within their songs. For instance, Kendrick Lamar's producer Soundwave used a famous Loopmasters sample from Splice on his 2017 album *Damn* (available until April 2017) called "COF_125_Am_LaidOut_Underwater.wav."[5]

However, royalty-free sample libraries predate the existence of CBMP services by many years. The main innovative feature of services such as Splice, Noiiz, and Loopcloud is that they aggregate numerous royalty-free sample libraries using centralized digital databases. By collecting these resources, CBMP platforms make sample-based music production much more efficient. For instance, Big Fish Audio has provided beat makers with royalty-free sample libraries since 1986.[6] Today, CBMP services provide streamlined access to samples that labels such as Big Fish Audio create.

While providing users with easy access to audio samples is their core service, many CBMP platforms also provide plugins, such as synthesizers, compressors, or EQs, as well as synthesizer presets. Splice and Noiiz, for instance, provides users with patches for Native Instruments' *Massive*, Xfer Records' *Serum*, Lennar Digital's *Sylenth*, and others. Thus, CBMP services provide users with access to networks of sounds, sound generators, and sound-shaping devices.

0.3 SOUND-TAGGING AND SEARCH FILTERS

To aid searchability, CBMP services *tag* samples and presets with descriptive language, which users can use to filter search results (Figure 0.2). For

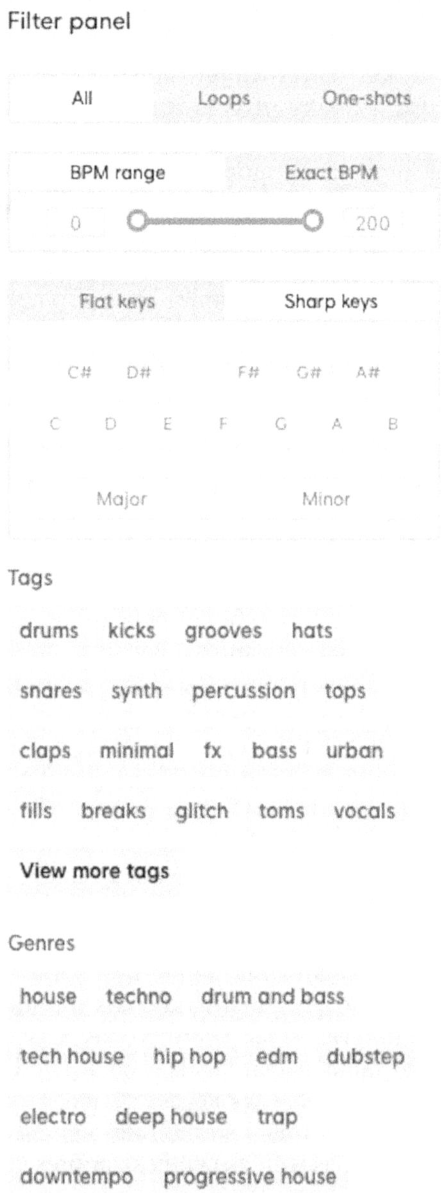

Figure 0.2 Illustrates Splice search filters.

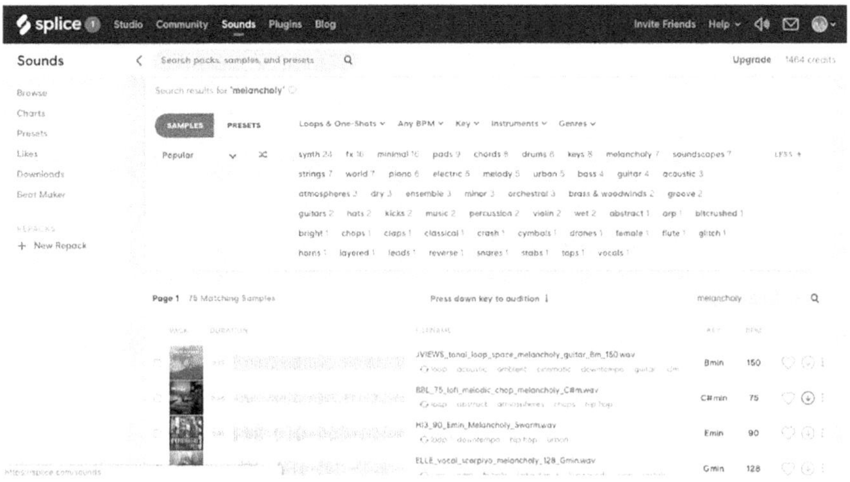

Figure 0.3 Demonstrates the way that Splice handles abstract search terminology.

instance, if I use the search term "kick drum," I will receive results which I can organize by genre, sample length (i.e., a loop or a one-shot sample), sonic description (i.e., distorted, acoustic, synthesized, noise-based, tonal, and others), bpm, genre, and key. Search results can be ordered by popularity, relevance, most recent, or via an entirely *randomized algorithm*. Depending on whether producers are seeking a unique sound, a more well-known sound, or something they perhaps have never heard before, they can select the corresponding option.

The search engine also accepts more abstract terminology (Figure 0.3). For instance, if I search "melancholy" on Splice, it provides 75 sample results (and 11 sample packs).[7] From this point, I can sort through the results by applying various filters. For instance, if I click the filter labelled "pads," the number of search results is reduced to 8. Thus, these CBMP services significantly reduce the amount of time it takes to find the right sounds for any project.

By incorporating these search features into centralized libraries, CBMP services have alleviated a major pain point for music producers. In the past, those who produce beats might sign up for several subscription services through Loopmasters or *Computer Music Magazine*, or they would purchase individual samples or packs through one of many available loop-production services such as Loopmasters, Capsun, and Bingoshakerz. These sample packs were costly, and, in addition, they required producers to constantly sift through numerous web-based sources. While this digitally mimics the tradition of *crate-digging* in vintage hip-hop production—or the act of searching through old vinyl releases for one-shots and loops—it is also an expensive and time-consuming activity for producers. CBMP services greatly reduce the cost associated with purchasing royalty-free samples, as well as the time it takes to find them.

0.4 PROJECT BACKUP AND SHARING

In recent years, it has become a standard practice for studios to keep many terabytes of extra hard drive space available for creating backup copies and to safely store old projects. However, some CBMP platforms have disrupted how producers store these contingency copies of their work. This alleviates a significant stress which producers routinely face, and, as a result, most producers welcome the chance to remove the threat of losing their work.

Some CBMP services offer free and unlimited automatic backup services. Splice, for example, provides the following explanation of their *Splice-sync* feature as follows:[8]

> There are two options for how the app syncs your work, which you can change in the app's preferences page:
>
> 1. (default option) Sync projects in the Splice folder: After installing, a Splice folder and shortcut will automatically be created. Save a project into the Splice folder and the app will automatically sync it. You can also drag and drop existing projects into your Splice Folder to sync them.
> 2. * Sync projects anywhere: If you save a project anywhere on your computer or external drive, Splice will ask if you want to keep it synced with the cloud. You can change the sync setting for a project in the app preferences screen.

Once this backup system is in place, Ableton, Logic, Garageband, and Fruity Loops project files will be automatically uploaded to a remote server every time the user saves their work (Figure 0.4). Alternatively, Loopcloud allows producers to back up their library of audio samples so

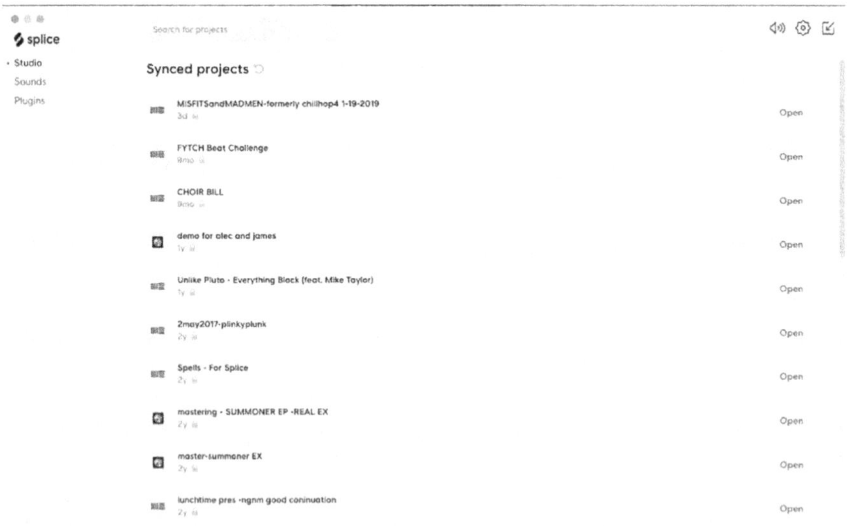

Figure 0.4 Demonstrates the Splice-Sync backup feature in Splice.

that they never have to worry that their creative materials—namely audio samples—will be lost due to a failing computer.

0.5 ACCESS TO PROFESSIONAL SOUNDS

CBMP services make professionally created sounds available to users for a low monthly fee. By using these services, even beginner music producers can familiarize themselves with the types of sonic resources heard on records they listen to. This considerably reduces the learning curve for those seeking to develop production competencies such as sample manipulation, synthesizer programming, editing, and mixing audio. In fact, users can access Just Blaze's drum samples (Jay-Z, Young Guru, Kanye West, Saigon, Eminem, The Game) which are featured on numerous chart-topping hits between 2000–2005 in a kit known as *Meow the Drums* (available on Splice). These samples were taken directly from his AKAI MPC, according to Blaze. In addition, producers such as Zaytoven (Gucci Mane, Migos, Travis Scott), Lex Luger (Gucci Mane, Wakka Flakka, Pusha T, Rick Ross, Juicy J), Getter (OWSLA, Shred Collective, Monstercat), Ducko McFli (Sremmurdrae), and FKi (Post Malone) have also released sample packs on CBMP platforms (Figure 0.5).

While some sample packs are created by famous producers, others are created by lesser-known (or unknown) producers and sound designers. These music professionals, who comprise the majority of contributions to CBMP libraries, create sample packs and synthesizer presets which capture the sonic zeitgeist of existing music scenes (such as hip-hop, trap, house, tech-house, techno, and others). Some of their contributions overtly emulate the sounds of well-known artists. For instance, one sample called "Bpm86_G#_36Chambers_EthnicPluck.wav" clearly draws inspiration from Wu-Tang Clan's cherished album from 1993, *36 Chambers*.

I encourage readers to search for some of the samples from Table 0.1, which draw inspiration from famous producers.

Figure 0.5 Shows some of the sample packs made available by famous musicians such as Just Blaze, Sonny Digital, Lex Luger, Boi-1DA, and Big Head.

Table 0.1 CBMP Samples Inspired by Famous Producers

Influential Producer	*Sample Name*	*Type of Sound*	*Sample Pack*
J Dilla	Bpm85_G_DillaDreamz_Drums03.wav	Drum Loop (continuous)	Shaolin Beats by Prime Loops
J Dilla	cw2_drm90_dillafly_full.wav	Drum Loop (continuous)	Chillwave 2 by Sample Magic
J Dilla	DKVPZ_clap_one_shot_dilla.wav	Snare Sample (one-shot)	Culture Alchemy by Rare Percussion
SwuM	SWUM_percussion_loop_03_150.wav	Drum Loop (continuous)	SwuM Drumkit by Splice (created by SwuM)
Wu Tang/RZA	Bpm86_G#_36Chambers_Snare.wav	Snare Sample (one-shot)	Shaolin Beats by Prime Loops
Just Blaze	JUST_kick_doctor.wav	Kick Sample (onc-shot)	Just Blaze—Meow the Drums by Splice (created by Just Blaze)
Lex Luger	LEX_LUGER_clap_trap_slap.wav	Clap Sample (one-shot)	Lex Luger—the Sequel by Splice (created by Lex Luger)
Ducko McFli	Ducko_percussion_loop_boujee.wav	Percussion Loop (continuous)	Ducko McFli Drum Essentials
Boi-1da	BOI1DA_808_08_C	Roland 808-inspired Sub Bass (one-shot)	Boi-1da Soundkit: Bare Sounds for Your Headtop by Splice (created by Boi-1da)
Sonny Digital	SONNY_D_hihat_closed_10.wav	Hi Hat Sample (one-shot)	Sonny Digital Drumkit

0.6 FREE TOOLS/RENT-TO-OWN

Free and inexpensive tools draw many users to CBMP platforms. At the time of writing, for instance, Splice offers users 82 instrument and effects plugins as a membership perk. These plugins come from companies including Xfer Records, iZotope, Valhalla, Voxengo, Plugin Alliance, and many others.

The types of free plugins Splice offers varies widely in selection, from signal processing effects to meters and synthesizers. Similar to samples and presets, free plugins are organized in a searchable database, and users can refine their search results using any of the following *tags*: utility, analog, imaging, modulators, delay, distortion, dynamic, filters, vintage, algorithmic, chorus, EQ, mid side, reverb, stereo field, synthesizer, amp simulator, bit-crusher, mastering, metering. If a producer wanted to find a free vu meter plugin to use on the master bus, for instance, s/he could type in "VU" and filter the results using the search tags "master" and/or "meter."

An advantage to this type of database is that it also allows users to search using tags, even in the absence of user-inputted text. For instance, if someone knows s/he wants the organ sample in a song to sound like it came from a vinyl record but s/he doesn't know exactly what type of plugin to use, s/he can simply click on the search tag for "vintage" or "analog." From this point, one only needs to try the different plugins until the satisfactory sound is achieved.

Another feature of some CBMP services is that they allow users to *rent-to-own* a variety of expensive DAW plugins. Splice, for example, allows users to lease popular synthesizers such as Serum, as well as mixing and mastering VSTs such as Ozone 8 and Neutron 2 ($249). As a result, producers can use these plugins for a small fee ($9.99/month) until the price of the plugin has been paid, at which point payments cease and the user *owns* his/her own copy. Splice plans to grow their plugin rental service in the future (Deahl 2017).

iZotope's Ozone 8, for instance, has a retail price of $499, which may prohibit numerous amateur producers from purchasing the software. However, with more affordable monthly payments, these producers can access iZotope's numerous mastering capabilities, including applications for spatial manipulation, tape emulation, multiband compression, linear phase EQ, and numerous others.

Xfer Records' Serum ($189) is an ultra-high-quality wavetable synthesizer, with an innovative animated graphic interface which allows users to modify wavetables in real time. According to the creators, they have gone to great lengths to ensure that the software is optimized using SSE2 instructions—the standard processing architecture for current Intel chips—in order to provide sophisticated, yet clean-sounding, oscillators.

Upon its release, Serum quickly became a popular plugin with tastemakers and musicians. Both *Electronic Musician* (2014) and *Computer Music* (2015) gave Serum editor's choice awards when it was released, for example. The plugin is also used by numerous famous producers, such as Kanye West, who was caught using a pirated version of the software in 2016 (The Guardian [online] 2016).

0.7 COMMUNITY, CONTESTS, AND COLLABORATION

CBMP services allow digital producers to share their projects with other users, find collaborators, and participate in contests. Users can upload DAW project files from Ableton, Logic, Fruity Loops, and Garageband into a searchable database, and anyone with a Splice membership can access these files. In addition, each database entry specifies the uploader's chosen plugins so that others can determine whether or not they have the corresponding applications. If a user does not have the plugins required for a project, s/he can either purchase (or lease) them through the links provided, or s/he may choose a substitute plugin. S/he can also use the Splice service to message the user and ask for an audio recording of the track in question (rather than MIDI, which is a set of instructions read by a plugin describing pitch, timing, velocity, and related parameters).

These file-sharing capabilities significantly reduce the strain of searching for musical collaborators, which can be a difficult task for not only hip-hop producers but musicians of all types. CBMP community databases allow users connect with one another based on genre type, bpm speed (which is a significant component of many digitally produced genres), DAW type, and a host of custom hashtags. And, users can easily preview projects in the database, making it easier to sort through them (Figure 0.6).

In today's music-making scene, CBMP communities provide a digital space for jamming, sharing, and releasing music. Interestingly, there are no limitations on the types of digital sounds, or music genres, heard within these communities. Although classical music receives little representation on CBMP services, for instance, synthesizer presets and audio samples which reference the genre are prevalent. In particular, CBMP services offer rich possibilities for those who create film, tv, or video game scores.[9]

For many digital music producers, remix competitions provide a chance to gain exposure. Quite often, remix competitions are hosted by established artists or labels for the purpose of promoting their work to other producers, DJs, and tastemakers. While dedicated websites have promoted these events for quite some time, CBMP services also facilitate remix competitions through the provision of tools for community engagement.

For instance, Splice hosts numerous mixing and remixing contests. Some of these contests involve creating new music from a sample pack, such as Brenmar's competition in the Summer of 2017, and some are remixing contests, such as Monstercat's November 2017 competition. For Brenmar's competition, the top four submissions won the chance to collaborate with Brenmar on a cloud-enabled collaborative EP called *The Collab Collection* (2017), the first of its type, which debuted on his record label known as High End Times. Monstercat's competition, on the other hand, offered producers the chance to release a song through their label, as well as a free Roland DJ-808.

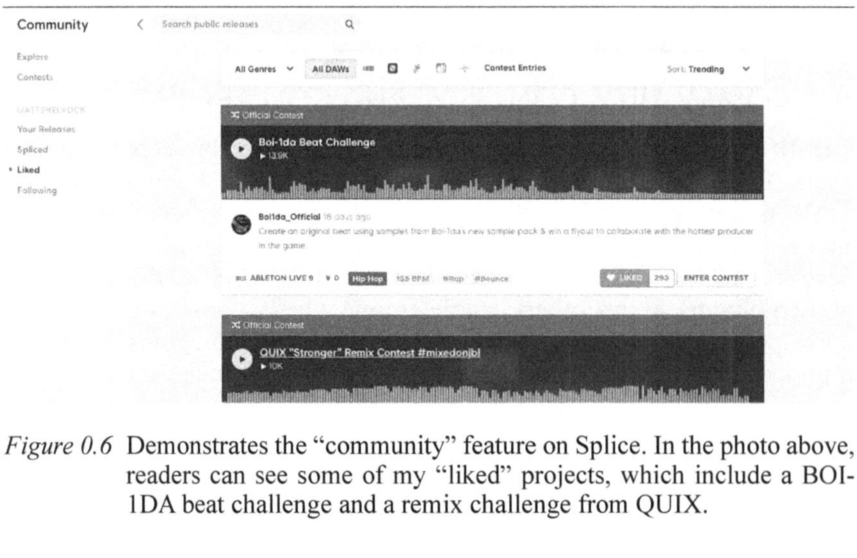

Figure 0.6 Demonstrates the "community" feature on Splice. In the photo above, readers can see some of my "liked" projects, which include a BOI-1DA beat challenge and a remix challenge from QUIX.

More recently, in spring of 2018, a multi-genre producer known as Fytch (bitbird, mau5trap, Dead Beats, Heroic) hosted a beat-making competition which invited users to make music using a folder of samples which he selected in advance. The winner and runner-up both received a private coaching session with Fytch on any topic related to music or music production.

These competitions provide producers with a chance to showcase their work within a larger music community. In many cases, producers who enter these competitions begin to follow one another's music and collaborate with each other. This provides value to these creatives since local amateur concert scenes rarely center around the creation of beats.

0.8 CHAPTER OUTLINE

Now that I have introduced readers to CBMP in a broad sense, Chapters 1–4 will provide a more granular level of information and analysis. In Chapter 1, I provide an overview of sampling and synthesis techniques, as well as an overview of the implications of connecting them with CBMP functionalities. In Chapter 2, I discuss mixing, which is another crucial aspect of beat-making supported by CBMP platforms, before turning the reader's attention towards the creative implications of using CBMP-based *royalty-free* samples (as they are commonly referred to) in music-making in Chapter 3. In this chapter, I provide a detailed explanation and analysis of how digital samples, and sample libraries, are used to produce hip-hop beats—a topic which is rarely explored in print media (if ever). This chapter will be the most useful for readers seeking to understand how samples are used to create new music.

In Chapter 4, after providing readers with some foundational background on mixing practices in Chapter 3, I explore the psychoacoustic dimensions of beat-making. I do so by elucidating some of the cognitive and psychophysical conditions which must be present in order for listeners to *believe in* a record, and I proceed to clarify how beat makers satisfy these conditions. In fact, CBMP services offer audio samples, presets, and mixing tools, which are all created for the purpose of aiding producers in establishing this *believability* on records. In Chapter 4, I explore how producers blend these sonic components to create records, which are, in essence, nothing more than sonic illusions.

The final chapter of this book covers a number of case studies in beat production. These case studies cover popular hip-hop styles including 1990s East Coast/boom bap, G-funk, Dirty South, lo-fi, trap, and experimental. Readers should know that, practically speaking, today's producers blend freely from these styles, as well as others. However, I believe that these categories are sufficient for demonstration, and it is my hope that these case studies inspire readers to engage in creative experimentation with sounds from these sub genres. It is my goal that these case studies will provide a basis from which other creative researchers can build upon.

NOTES

1. www.forbes.com/pictures/eeel45jfeg/1-tiesto-22-million/#6beb644e2cef
2. On many commercially successful hip-hop releases, a single executive producer (either an outside producer or the artist him/herself) will oversee the completion of the record. This individual is often responsible for facilitating recording sessions with emcees, but other responsibilities may include acquiring *beats* by other producers, hiring the required engineers/assistant personnel, or even performing engineering duties.
3. Please see https://splice.com/sounds/labels
4. Moreover, when he does take samples from records, he proceeds to "grab something, take a loop out of it, then flip it, reverse it, distort it, or chop it up. Then I ask myself 'Did I create something?'" (quoted in Pensado's Place 2014). In other words, it would be nearly impossible to identify the samples that Noah 40 uses to produce many hit rap, hip-hop, and R&B songs. When the samples are more easily identified, he simply opts to give rightsholders a portion of his publishing and master rights within his new work (while also paying any applicable fees).
5. This sample can be heard at the following web address: www.whosampled.com/sample/494321/Kendrick-Lamar-FEEL.-Loopmasters-COF-125-Am-LaidOut-Underwater/ accessed September 27, 2017.
6. www.bigfishaudio.com/about.html?1;24;1:::::::::::: accessed October 19, 2017.
7. Over the course of a few months, the number of search results for this more abstract term grew from 58 samples across 7 sample packs to 75 samples and 11 sample packs.
8. Taken from https://support.splice.com/hc/en-us/articles/201178607-Uploading-your-projects-to-Splice accessed December 1, 2017.
9. And, of course, hip-hop is a genre that samples audio from soundtrack sources often, such as Busta Rhymes' use of the *Psycho* (1960) theme in "Gimme Some More."

1

Understanding Samples, Synthesis, Editing, and Mixing

The previous chapter introduces some core cloud-based music production (CBMP) concepts. In this chapter, I cover a number of beat-making activities which are directly supported by CBMP services such as Splice, Loopcloud, and Noiiz. In particular, these services help producers streamline production activities such as (i) sampling, (ii) synthesis, (iii) triggering sounds, (iv) arranging, and (v) mixing. Most of the time, when these activities are discussed in literature, they tend to be treated as though they are completely separate from one another. And, in certain music production traditions, this may be true.

However, in most forms of digital music making, such as the creation of hip-hop beats, producers freely switch between composing, arranging, designing and recording sounds, and mixing. These production methods, and how they relate to one another, are the main focus of this chapter, rather than the classic DJ-centered techniques commonly associated with hip-hop in the 1980s and early 1990s. This is because most of today's beat producers prefer to work within DAWs, which make it quite easy to switch between beat-making activities such as sampling, synthesis, triggering sounds, arranging, and mixing (Shelvock 2017a: 172).

In some cases, such as Fytch's Spring 2018 hip-hop beat-making challenge on Splice, producers may also combine CBMP resources with live instruments. Yet, as Fytch demonstrates, when producers incorporate live instrumentalists within their work, they often treat these recordings like any other sample downloaded via a CBMP service. That is to say, through the use of editing technologies, they mutate the rhythmic and timbral properties of these performances in the same way as they do with pre-recorded samples provided by Splice, Loopcloud, or Noiiz.

Since the sounds producers create using digital production methods exert such a clear influence on hip-hop's key influential instrumentalists, in this chapter I focus on these computer-based music-making techniques. While a very small portion of hip-hop beat makers use live instruments exclusively (or at least mostly) within their work, this is somewhat of a unique approach. Groups such as The Roots or Bad Bad Not Good feature drummers, bassists, guitarists, keyboard players, and others who try to emulate the sound of hip-hop sampling when they play. In order to accomplish this, they simply draw inspiration from the type of rhythmic phrasing predominantly heard on quintessential sample-based hip-hop records. Indeed, sampling is such a foundational technique within this genre that even instrumentalists try to perform as though they have already been sampled and edited by a producer.

In the sections of this chapter, I provide a brief overview of the most crucial hip-hop production techniques so that readers can see how CBMP supports practices related to (i) sampling, (ii) synthesis, (iii) triggering sounds, (iv) arranging, and (v) mixing and how each of these (i-v) practices coalesce during the beat-making process. While additional approaches and techniques exist, I have chosen the methods which are most commonly discussed (or used) by producers in popular video series offered by Mass Appeal, Waves, Maschine, FL Studio, Ableton, HotNewHipHop, and others. In addition, I have chosen the techniques I encounter most often as a producer, manager, publisher, and label representative. However, I do not claim that any specific production technique I mention constitutes some form of secret knowledge, or that these tools are unknown to practitioners. In fact, I argue just the opposite: these production methods are used on a routine basis by beat makers, yet no peer-reviewed text provides an in-depth analysis of how these techniques work together to create hip-hop beats *per se* (although many authors do, of course, discuss topics surrounding beat production through the lens of cultural theory, such as Schloss 2014). It is my hope that what follows will spark increased discussion on the topic of record production, as an area worth analyzing in-and-of itself, so that many of the misunderstandings which currently pervade popular music texts can finally be laid to rest.

1.1 VIRTUAL PERFORMANCE: PERFORMING SAMPLES, PERFORMING SYNTHESIS

In order to make music using audio samples and synthesizers, producers record themselves while triggering these sounds using a variety of tools. Today, this almost always occurs through the use of a DAW, such as Ableton or Maschine. This section describes how producers incorporate various performance technologies, such as MIDI keyboards and trigger pads, in order to create hip-hop beats.

1.1.1 Performance Inputs: Sample-Triggering, Synthesis, and MIDI

One particularly important tool for hip-hop beat makers are sample-triggering devices, such as the classic instruments offered by Roland or Akai. In addition, today many producers have adopted new tools offered by companies such as Novation, M-Audio, or Native Instruments in order to accomplish the same task as these well-known machines. One shared feature of many of these hardware interfaces is the provision of a series of square buttons pads, which are arranged in a grid format. These pads can trigger samples, but they can also control software synthesizers in order to produce melodies because they transmit data which software synthesizers can transform into discrete pitches. Some controllers, such as the Novation Launchkey, include both a series of button pads, as well as a piano keyboard, and users are free to switch back and forth between them in order to control or trigger sounds (Figures 1.1 and 1.2).

Most of these devices make use of a form of musical software data known as MIDI, which stands for Musical Instrument Digital Interface. MIDI is a form of musical data which is stored by computer software, such as the samplers discussed above or DAWs, which contains metadata describing a number of standard musical parameters. This data is stored using software and can be recalled at any time. Some of the musical parameters which MIDI controls include pitch, velocity (i.e., dynamics), note length, and rhythm.

MIDI allows users to store and modify this musical information by either (i) using controllers to record virtual performances with sampler devices and synthesizers or (ii) modifying MIDI data outside of real time.[1] In the first case, producers can simply record the performance data which enters into the computer via a controller such as NI's Maschine or Novation's Launchpad. Alternatively, producers can also modify or create MIDI data outside of a real-time performance. They may edit pitches, rhythms, dynamics, and other sonic parameters by prescribing these things while

Figure 1.1 A Korg ES1 sampler has a tube preamp which can add additional warmth to digital samples triggered on this unit. This device can also act as a simple MIDI trigger device.

Figure 1.2 A Novation Launchkey provides users with 16 square-shaped trigger pads which correspond to various MIDI notes. This device also has a two octave MIDI keyboard. In addition, the 8 dials on the top can be assigned to various DAW controls using its CCN function.

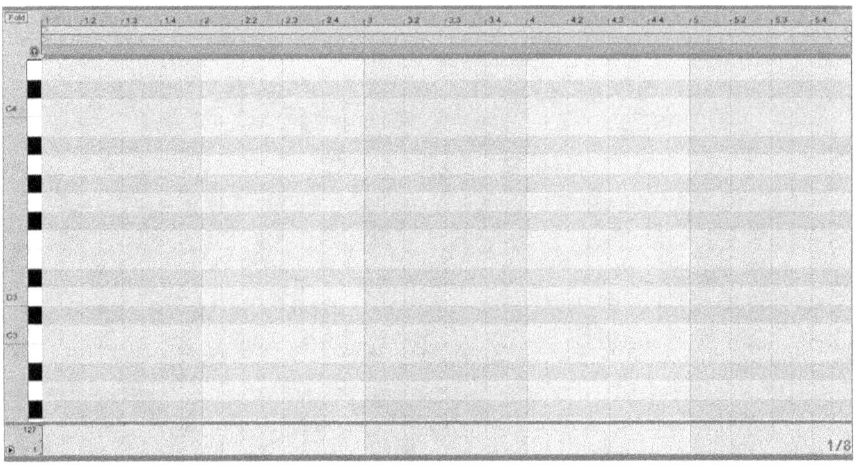

Figure 1.3 Ableton's matrix editor is used for altering MIDI events.

the producer is not performing. To do so, they rely on a user interface known as a *matrix editor*, which allows them to input exact note lengths, timings, pitches, and velocities (i.e., dynamics). When producers edit or create MIDI data within a matrix editor, it resembles the act of score creation (Figure 1.3).

MIDI also allows user to store a second type of controller data, known as continuous controller number (CCN) data, which can store and play back pitch wheel modifications on a keyboard, for example. An even more pertinent feature for those who use these controllers includes using CCN data to control a variety of effect parameter knobs, such as filters, standard EQs, and other effect parameters, by assigning CCN values to various software parameters.

The CCN functions on MIDI controllers are fully assignable. When CCN data is used in tangent with an interface for inputting notes and rhythms, such as a keyboard or a matrix of square-shaped triggers, producers can control almost any conceivable musical (i.e., notes, rhythms, pitches, velocity, and note duration) or sonic (i.e., frequency balance, dynamic contour, ambient profile, stereo configuration) property in real time. Typically, these hardware interfaces have a series of rotary knobs, sliders, and buttons which users can program to configure various effects such as EQs, compressors, filters, ambient effects, and modulation-based processors. For example, a user can assign the Q-value for one filter on a parametric EQ to an empty CCN channel. This CCN channel can then be assigned to a dial on a MIDI controller. From this point, users may freely adjust the Q-value on this filter by simply turning the associated dial.

Typically, producers use CCN data for two purposes. These reasons include (i) controlling sampler or synthesizer parameters and (ii) specifying signal processing parameters. In both cases, CCN data facilitates a more tactile approach for altering both software instrument parameters and effects in real time in comparison to keyboard and mouse input. Another important advantage to this approach is that it allows users to record the signal as its sonic parameters change over time using a process known as

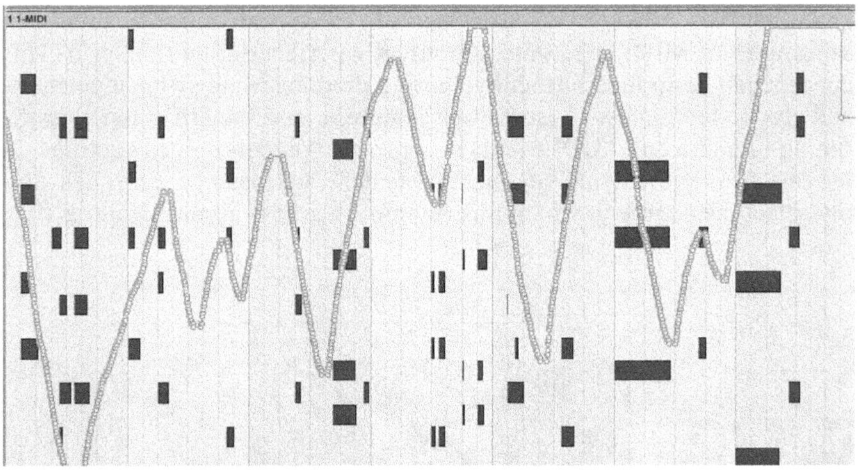

Figure 1.4 MIDI automation data in Ableton. Ableton records automation data and makes this available to the user in the so-called automation lane. Conveniently, this data is presented chronologically. It also overlaps the visual representation of the track's MIDI data, which is very useful for editing. This feature helps users aurally relate parameter automation events to various pitch sequences.

automation (Figure 1.4). A producer could, for example, cause the wet/dry setting on a phaser effect to increase and decrease at specific times over the course of a project. Of course, many users also prefer to program these types of modifications using keyboards and mice, which is known as the *point-and-click* method by practitioners. However, the lion's share of producers freely switch between both modes of operation.

When using MIDI to control synthesizer parameters, producers may assign available CCN channels to control a variety of settings. To return to Massive, as an example, any of the software's parameters, such as those regulating oscillators, filters, amplifiers, and modulation (all of which are explained in more detail later in this chapter), may be controlled using an interface capable of broadcasting and receiving CCN data. For instance, software such as Massive and Serum also makes it possible to control groups of synthesizer parameters at one time using the software's 8 *macro* controls. Users can modify 1 or more settings simultaneously by assigning them to these controls. From this point, a DAW can track, or record, all of the changes made in real time.

1.1.2 Point-and-Click Editing

As seen in the screenshots in the previous section, all of the MIDI data captured by the DAW is easily accessible. This allows users to modify MIDI data by using their mouse and keyboard. Given the fact that MIDI and CCN data can control nearly any musical parameter within a project, producers can use their DAW's editing functions in order to exert a high level of control over a number of minute musical details.

Users can also engage in algorithmic editing by applying various quantization settings. Most DAWs allow users to easily snap music events, such as samples or MIDI data, onto a metrically perfect digital time grid. This concept can be applied further by altering rhythms using swing algorithms and groove templates. These editing features cause the DAW the alter the timing of audio and MIDI events according to predetermined rhythms. In the case of Logic's swing quantization feature, for instance, users can simply select the degree to which this rhythmic effect is applied (Figure 1.5).

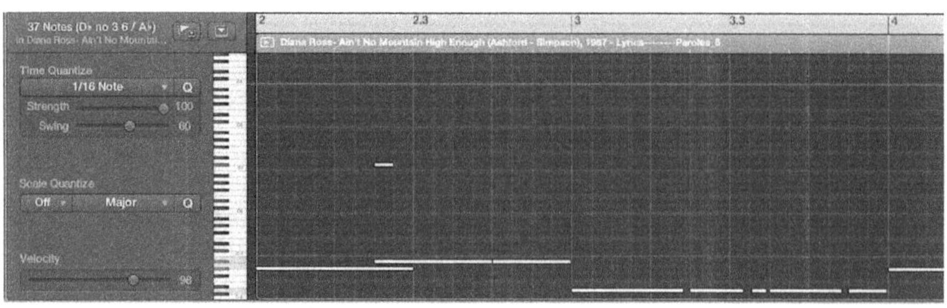

Figure 1.5 Quantize and Swing in Logic X. These features can be used to alter the timing of MIDI events.

Figure 1.6 Ableton's groove pool feature. The groove pool allows users to import various rhythms, phrases, and timing nuances into a project and then apply these to various tracks in the project. This causes the track's timing events to be altered in conjunction with the data contained in the groove pool.

In Ableton, users can apply groove templates in order to alter the rhythms of audio and MIDI. Ableton comes with a variety of preset grooves which can apply varying amounts of swing to a performance, as well as well-known longer grooves such as bossa nova rhythms. Users can also add to their collection of groove templates by simply selecting any audio or MIDI clip and then "extracting" its groove features. Once it is turned into a template, the producer can apply the groove to varying degrees on a percentage basis (Figure 1.6).

1.2 CREATING HIP-HOP SOUNDS: SAMPLING

As it stands, many musicians believe that sampling merely involves choosing one or two bars of pre-recorded music and then continuously looping this selection in order to create an instrumental track. But, as mentioned in this book's earlier chapters, this does not accurately represent the type of sampling practice that most professional and amateur producers actually use to create music. In some cases, authors have provided a post hoc analysis of records which use sampling (Sewel 2013, 2014); however, few (if any) sources provide a direct analysis of the creative acts around record creation in this genre.

In music production, sampling involves making new music from pre-recorded audio. Samples are used in conjunction with a DAW's editing and arrangement features in order to perform a variety of compositional functions. For example, in hip-hop, audio samples may be arranged in such a way that they comprise a song's percussive instrumentation, as well as its melodic and harmonic features.

In fact, almost every instrumental in this genre incorporates audio samples of some type (Shelvock 2017a: 173). And, many of the samples which producers use to create hip-hop beats resist simple categorization as *melody* or *harmony*. Despite this fact, sound effects (sometimes referred to as *sound FX*), oblique sonic textures, and various atmospheric samples are nonetheless crucial to hip-hop's musical language. These sounds are easily found, on a license-free basis, within popular CBMP platforms such as Splice, Loopcloud, and Noiiz.

In order to make music from audio samples, one has to (i) upload them to some type of *sample-trigger* device which allows the producer to *perform* with them or (ii) cut, paste, and edit samples using the sample-arrangement window within a DAW using the *point-and-click* method. The popular YouTube series *Rhythm Roulette* celebrates the first method by showcasing hundreds of producers who create music using samples which they trigger using MPCs, Maschine, Novation Launchpads, various keyboards, and related instruments. However, when a producer uses the sample-triggering method, s/he typically edits his/her performances to some degree by using a point-and-click approach after the fact.

In CBMP, the samples which producers use are already free of any license restrictions. However, in the past, samples were taken from previously released commercial recordings without permission. This approach always impedes on the intellectual property of other musicians. To this day, sometimes producers are taken to court for alleged copyright infringement.[2]

Depending on the nature of the sampling techniques used, and the source that musicians sampled, many sample-based producers did not face litigation throughout the golden age of hip-hop (c. 1980s–1990s). This is because, in many cases, musicians acquired licenses to use samples through legal channels, known as sample clearance. Musicians who did not follow sample-clearance protocols would often utilize shorter samples (i.e., less than 1 second) which were much harder to trace to the original source. And, while these short samples are not legally permissible, it is quite difficult to prove the origin of these stolen sound bytes.

These short-form samples, whether acquired legally or illegally, are known as *one-shots*.[3] One-shot samples are small audio files that typically contain a kick drum or snare drum being played only one time, for instance. Producers typically use these one-shots in conjunction with a sequencer (or DAW) to create a song's percussive instrumentation.

Producers may also create short samples from any musical sound in order to create new musical phrases. In addition, one-shots are commonly reversed, edited, or re-pitched in a number of ways before being loaded into some type of sample-triggering device, which I discuss below. Noah "Shebib" 40 (OVO Sound, Drake, Lil Wayne, Action Bronson) remarks on his own use of samples, stating:

> I use em sort of as a tool I guess, you know? I'm a fan of publishing, so I try not to sample *too* much, but it's a tool in your arsenal. There is something to be said about not sampling, and if you

> say "I'm not going to take *that* from someone else," I say "great, congratulations—I'm happy for you." But, you *can* sample, and what if a sample makes the song better? What I try to do is grab something, take a loop out of it, then flip it, reverse it, distort it, or chop it up. Then I ask myself "Did I create something?"
>
> (Quoted in Pensado's Place 2014)

Sample-based techniques are even more prominent today than they were throughout the 1980s and 1990s due, in part, to the overwhelming prevalence of what are commonly known as "royalty-free" samples to music producers. In more precise terms, these are samples which are free of all licensing restrictions. And, of course, these are the types of samples which CBMP services make available *en masse*.

While most available scholarship discusses sampling as an act of social resistance, this is quickly changing as cloud-based sample databases become more popular. When Splice launched, it featured over 500,000 samples from 1,500 unique sample packs. Now, 6 years later, its collection has grown to over 2 million samples. A number of specialized recording labels now exclusively make recordings for producers to sample. Splice, for instance, has signed deals with 120 of these companies. Some prominent examples include Loopmasters, CAPSUN Pro Audio, and Class A Samples.

1.2.1 Technological Constraints, Sample Arrangement, and Sample Phrasing

In sample-based music, several variables constrain the creative choices one can make during the production process. And, these limitations impact the final configuration of all sample-based music. I use the term *sample arrangement* to describe (i) the act of composing music from samples, as well as (ii) the sonic results of using samples within a mix. In the same way, musicians often discuss the act of musical arrangement as an activity, while a song's arrangement refers to the design configuration of its compositional elements.

Simply put, *sample arrangement* occurs when a producer configures samples for use within a music project. In addition, a track's *sample arrangement* can also refer to the musical artefacts which are created via the use of samples. These can include rhythms, harmonies, melodies, effects, and atmospheric textures. I use the term *sample phrasing* for a similar purpose in order to describe both (i) how one structures rhythms within a sample-based beat and (ii) the rhythmic results of using samples to create beats.

In order for any sample to simply exist, someone must first digitally record a sound. Digital recordings work by capturing and storing thousands of short samples of a sound as one long chain, which can be recalled, or played back. Each of these samples within the chain encodes the data required to reproduce the original waveform at discrete points in time.

Only when these samples are played back in rapid succession at the correct sample rate—namely, either 44.1 or 96kHz—the listener hears an uninterrupted sonic facsimile of the audio which was recorded. When these conditions are satisfied, the audio is essentially re-animated for the listener in a way that resembles stop motion photography (Hodgson 2014: 90). Assuming the recorder's sample rate is set correctly, in order for a sound to appear on a recording, it must also have (i) a frequency distribution which can be recorded (i.e., 20Hz–20kHz) as well as (ii) an acceptable dynamic range (i.e., 96–120dB).

When each of these technical conditions are satisfied, one can begin using a sample with a sample-triggering device of some type. Most often, these devices are softwares, such as Maschine (NI), Sampler from Ableton, or the EXS in Logic, and producers use them in conjunction with hardware devices. These devices feature a matrix of buttons, or a keyboard interface, on which users can assign one or more samples. When these buttons or keys are played, the device triggers the corresponding sample. When one sample is used multiple times across numerous keys or pads, producers often configure the pitch so that it increases (or decreases) as they progress stepwise through the buttons on their chosen sample-triggering technology.

These sample-trigger applications also constrain one's creative choices. Sample-triggers typically rely on MIDI data. If a user loads an audio sample on to each available key or button on a sample-trigger, each of these will also have a MIDI assignment containing an octave number (0–10) and a note name (all 12 Western pitches: A, A#, B, C, C#, D, D#, E, F, F#, G, G# or their enharmonic equivalents). MIDI assignments are limited to 127 values from C0–G10.

When using MIDI data, users are constrained by the types of musical parameters MIDI can record: note (i.e., C0–G10), note duration, timing, and velocity (i.e., the intensity at which the trigger was played). In addition, producers can control the behaviour of the sample, and the sample-triggering software, using continuous controller number (CCN) data. For instance, a producer might increase or decrease the length of a sample by assigning it to a hardware device's CCN to shorten or lengthen it according to the user's input.

In addition, once a producer uses a sample, s/he is also creatively constrained to the sound characteristics of the chosen sample. No sampling technique can cause a recording of a guitar, for example, to magically transform into a recording of Werner Herzog reciting the alphabet. So, once a producer chooses a sample to work with, the end result is constrained by the sound of the sample—even if a producer causes it to sound much different than it initially did.

In a material sense, the above technological limitations determine the musical phrases producers can create using samples. However, producers do not necessarily *feel* this constraint any more than a professional vocalist feels limited by having to use his/her own voice. In the same way, a professional drummer rarely perceives any sort of constraint when using an ideal drum set, even though his/her performance can only include the drums which make up his/her kit. All musicians perform within boundaries of some type.

Simply put, the terms *sample phrasing* and *sample arrangement* take into consideration the fact that producers use samples in an intentional way, rather than a seemingly random or irrational way, based on the capacities of the technology they use. Like with other music-making devices, performances which rely on sample-triggering devices are creatively constrained by the instrument itself. In sample-based music making, this includes (i) the samples chosen for triggering, (ii) the capacities of sample-triggering software itself, and (iii) the capacities of the hardware device used to communicate with the sample-triggering software. I explore the sample phrasing in more depth in the next chapter.

1.2.2 Types of Samples: One-Shots and Continuous Samples

There are two main categories of samples which producers use: one-shot samples and continuous samples. Each of these types favour different sample phrasing strategies. One-shots are short samples which play intermittently throughout a track, and they rarely last for more than one second. These shorter segments, sometimes called slices by producers, are typically configured into sequences which sound like performances.

Audio Example 1.1

In Audio Example 1.1, I demonstrate an assortment of one-shot samples played one after another.

Continuous samples instead play for a longer duration, and they often contain longer performances originally created by other musicians or loop-makers. Many of these samples are similar to so-called *breaks* used in classic hip-hop. Famously, producers in the classic hip-hop era would sample drum breaks from popular songs and repackage these into new recordings which also included emceeing and a variety of other sounds. Sample creators such as Loopmasters imitate the sounds of these breaks and make them available in sample packs, which are free of licensing restrictions (as heard in *RawCutz* and many others). CBMP services collect sample packs like these and make them easily searchable by users.

Audio Example 1.2

In Audio Example 1.2, I demonstrate an assortment of continuous samples (i.e., breaks) played one after another.

Drum samples are the most commonly used type of one-shot heard in hip-hop beats. The libraries which CBMP services offer contain drums sounds with nearly any conceivable timbre, velocity, or playing style. Users can apply search filters to a query, such as "snare drum," for example, in order to find a sample which suits their creative needs. For instance, one might want to use a warm 1970s snare drum, similar to the one heard in the intro of Stevie Wonder's "Superstition" (1972). In this case, CBMP libraries make it easy to add filters such as "warm," "analog," "funky," "vintage," or other semantic descriptors.

Audio Example 1.3

In Audio Example 1.3, listeners will hear a variety of hip-hop snare sounds which fit semantic descriptions such as "warm," "analog," "funky," and "vintage."

Producers typically control one-shots using a physical controller of some type. They load a number of these samples into a software or hardware device known as a sampler, which allows them to trigger these sounds using a control surface. By doing so, beat makers effectively turn sounds into instruments. Most producers use sample controllers which integrate with MIDI technology by assigning various audio files to each of the 127 available MIDI notes. Each time the producer plays 1 of these notes, the sampler triggers the corresponding audio segment (Figure 1.7).

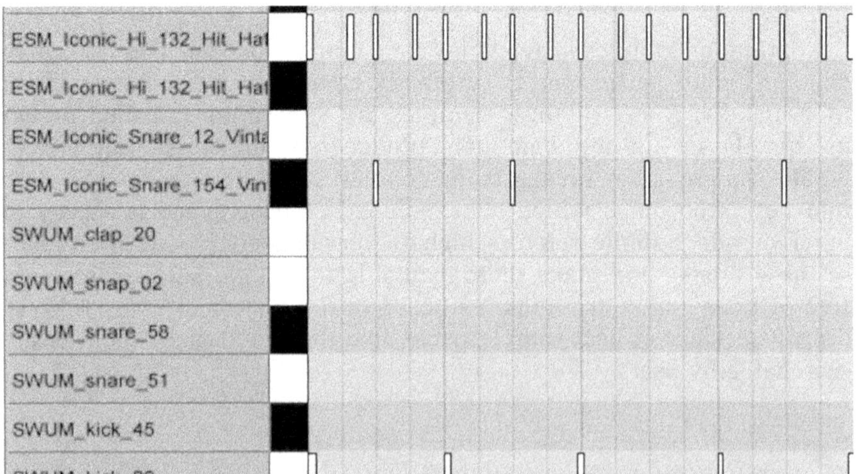

Figure 1.7 Beat from one-shots. Demonstrates the corresponding MIDI data for Audio Example 1.4. This is a drumbeat made from one-shots.

Audio Example 1.4

In Audio Example 1.4, I arrange a number of one-shots together to make a hip-hop drumbeat in the style of J Dilla. The MIDI data which corresponds to this drumbeat is shown in Figure 1.7.

While most of the one-shot examples above include percussion sounds, this is not a categorical restriction. One-shots may also consist of sound effects, melodic sounds, harmonies, and atmospheric textures. Keyboard-based controllers are quite useful for controlling these sounds; however, it is just as common for producers to instead use a button controller, such as NI's Maschine or Novation's Launchpad.

Audio Example 1.5

In Audio Example 1.5, I demonstrate a variety of melodic one-shots, as well as a number of FX-based one-shots. While many one-shots are rhythmic sounds, many others are not.

Beat makers may also turn continuous samples, also known as loops, into shorter-duration one-shots. Since CBMP samples have very few licensing restrictions, producers edit and mutate continuous samples into shorter samples for the purpose of creating new sounds. This type of sound-designing process is integral to the creation of many hip-hop beats (Figure 1.8).

Audio Example 1.6

In Audio Example 1.6, I turn a simple continuous loop containing some chords played on a Wurlitzer into a new composition. First, I demonstrate the original (i.e., untouched) audio sample, which contains four chords. Then, I assign each of these chords to a discrete pad on my MIDI controller. At this point, I re-create the original audio sample by playing each pad in sequence (i.e., pad 1, pad 2, pad 3, pad 4—each of which corresponds to the four chords in the original audio sample).

Once this occurs, listeners will hear my new creation, which I have paired with a drum sample called 80_BoomBapBeats_3_TL.wav (from *Hip-hop Break Ups* by Touch Loops). Listeners will notice that the four chords from the original sample are rearranged, repeated, and take on a vastly different rhythmic qualities from the original version.

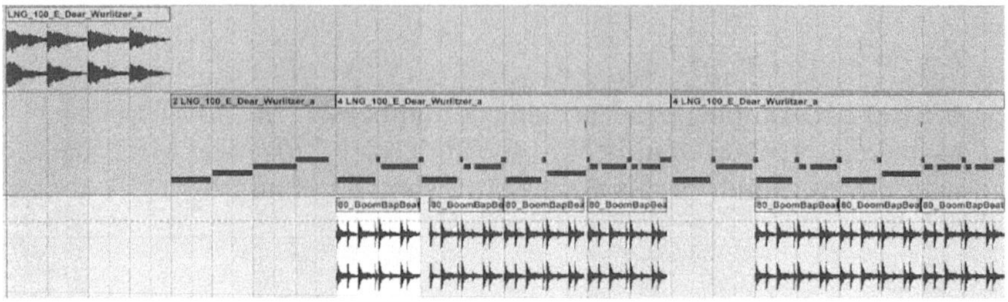

Figure 1.8 (i) A continuous audio sample (track 1, on the top left) which has been (ii) sampled and controlled via MIDI (track 2), and then (iii) paired with a drum sample (track 3, on the bottom).

When producers edit continuous samples into smaller slices, occasionally these segments contain rhythmic content which does not align perfectly with either (a) the groove of a project or (b) the digital time grid. Sometimes, a producer may opt to use a mixing technique, such as sidechain compression, in order to shoehorn two rhythmically conflicting sounds together (I discuss this more in this book's chapters on mixing). However, in many cases, producers might simply keep sounds wherein the timing does not match. In some cases, such as those highlighted in Table 1.1, imperfect rhythms such as this become a cherished feature of the record:

1.2.3 Timbre and Arrangement

A sample's overall timbral make-up constrains the way a producer can use it within a project. In this section, I discuss how the sonic characteristics of a sample impacts the producer's arrangement choices in hip-hop. In this regard, one criterion which producers may consider includes the sample's sonic resemblance to either (i) vintage or (ii) modern recordings. Even if a producer does not actively contemplate this decision, most samples and sounds in hip-hop production are easily categorized along these lines. The presence of one or both of these timbres within a project tends to have different implications for a project's arrangement characteristics. One additional criterion is the overall timbral density of a sample (i.e., dense sounds versus sparse sounds), which also constrains arrangement decisions in this type of music.

In hip-hop beat production, the timbre of various samples can be categorized along a spectrum, with (i) *vintage* sounds at the left side of the spectrum and (ii) *modern* sounds on the other (Figure 1.9). By *vintage*, I refer to samples which sound like they were created using older analog circuitry. Listeners tend to describe these sounds as *warm* and rich. By *modern*, I refer to samples which sound like they were created digitally. These sounds tend to exhibit a higher average amplitude than vintage sounds, as well as a more complete frequency distribution. In general, modern sounds tend to contain more spectral energy in the lowest audible frequencies (20–150Hz) and the highest audible frequencies (10–20kHz).

Table 1.1 Commercially Available Hip-Hop Tracks With Timing Misalignment

Song—Artist	*Timing*	*Auditory Cue*
"Suede"—NxWorries	00:02 (throughout)	Drums and bass: bass instrument, kick drum, and snare are behind the hi hat (which is faster)
"Sheets"—jhfly	00:07 (throughout)	Kick drum and hi hat: hi hat occurs ahead of kick drum. Hi hat also exhibits a slight swing.
"Goldie"—A$AP Rocky	00:00–00:27	Drum break sample (hi hat, kick drum, and snare) and flute: the flute sound rhythmically occurs ahead of the drums
"Realer than Most"—Closed Sessions, Mick Jenkins, Daily Auston, Noname, Saba, OnGaud	00:46 (throughout)	Kick drum and hi hat: kick drum rhythmically occurs behind the hi hat and snare
"Gold Soul Theory"—The Underachievers	00:03 (throughout)	Audio sample (arpeggio sequence): this melodic sound rhythmically occurs behind hi hat
"Ms. Fat Booty"—Mos Def	00:00 (throughout)	Bass instrument and kick drum: bass instrument rhythmically occurs after the kick drum, and both occur after the hi hat
"Mama Said Knock You Out"—LL Cool J	00:01 (throughout)	Bass instrument and kick drum: bass instrument rhythmically occurs after the kick drum, and both occur after the hi hat
"Juicy"—Notorious B.I.G.	00:22 (throughout)	Kick drum and toms occurs late in comparison with hi hat and snare
"Bonita Applebum"—A Tribe Called Quest	00:21 (throughout)	Bass and kick drum rhythmically occur later than hi hat and snare
"Tic Toc"—Lords of the Underground	00:06 (throughout)	Iconic snare drum occurs lightly behind the beat
"Fu-Gee-La"—The Fugees	00:11	Keyboard sample (i.e., mellotron) is rhythmically behind the drums

Audio Example 1.7

In Audio Example 1.7, listeners will hear a Wurlitzer performance which I have sampled and chopped. This keyboard instrument rhythmically occurs behind the other percussion instruments—especially the woodblock instrument, which enters at the 6-second mark.

Figure 1.9 This diagram illustrates the notion that hip-hop samples tend to favour a vintage or modern timbral design. Vintage samples contain similar spectral distributions or artefacts as records made using vintage equipment. Modern samples tend to have a cleaner, more "digital" characteristics. Spectral diagrams in the next chapter also reinforce this point.

Figure 1.10 The left side of the spectrum: vintage samples.

Many samples exhibit sonic characteristics which place them firmly at either end of the spectrum; however, a large number of samples also fit somewhere along the middle.

I use the term *vintage sounds* to describe any timbre which reminds listeners of the characteristics of vinyl records from the 1950s through roughly the 1980s (around the time when vinyl began to fall out of favour). More specifically, *vintage* in this case refers to sounds which tend to have a rich harmonic distortion or saturation characteristic present. In terms of frequency distribution, these sounds also typically feature an abundance of amplitude in the low-mid range (i.e., 250–500Hz) and middle range (i.e., 500Hz–1kHz). Vintage samples often contain less high frequency data (particularly above 10–15kHz), as well as less sub bass (especially below 80Hz). Modern sounds exhibit the opposite characteristics: they tend to sound *clean*, with less dominant low-middle and middle frequencies. Qualitatively speaking, when modern samples contain distortion, the distortion does not sound like it came from analog equipment, such as a tape machine with a high input level. Instead, it takes on a harsher characteristic, with more distortion present in the upper frequencies. Readers should note that, in the next chapter, I provide some exemplary spectral analyses of these different types of samples.

Since sampling vinyl records was so integral to the sound of classic hip-hop throughout the 1980s and 1990s, it is hardly surprising that vintage sounds are so popular on today's hip-hop records. Popular vintage-style sample packs such as *Meow the Drums* (Just Blaze), *Iconic Hip Hop Drums* (Epic Stock Media), *Iconic Vintage* (Epic Stock Media), *Hypercolor Hip Hop* (IQ Samples), *Vinyl and Tape Drum Machines* (Sample Magic), and others feature samples which sound noticeably *warm*. By this, I mean that many of these sounds have pleasant-sounding audible distortion, as well as an abundance of amplitude in the low-mid and middle frequencies. These types of sounds belong furthest to the left on the vintage-modern sample spectrum (Figure 1.10).

Audio Example 1.8

In Audio Example 1.8, I demonstrate a number of timbrally *warm* samples. Listeners may notice that these sounds closely resemble sounds used in many classic hip-hop tracks from the late 1980s through the 1990s.

Modern hip-hop samples are often louder and cleaner sounding than vintage ones. By this, I mean that modern samples tend to feature a higher average amplitude and tend to exhibit less distortion. In samples where distortion tends to be present, such as kick drums and software emulations of the 808 sub bass, the distortion tends to take on a harsher characteristic that one would typically hear from vintage equipment when it distorts. Listeners often describe this type of distortion as harsher sounding. Some popular sample packs which embrace this style include *Future Soul Melodic RnB and Trap* (Origin Sound), *Future Trap* (Prime Loops), *Lex Luger Drumkit* (Splice), *Lex Luger Sequel* (Splice), *UZ—Real Trap Samples Vol. 1* (Splice), *Future Trap by Singular Sounds* (Singular Sounds), and *Zaytoven Drumkit* (Splice). These types of sounds belong furthest to the right on the vintage-modern sample spectrum (Figure 1.11).

Audio Example 1.9

In Audio Example 1.9, I demonstrate a number of timbrally *modern* and clean-sounding samples. Listeners may notice that these sounds closely resemble sounds used in many hip-hop tracks in the early 2000s through today, particularly in trap music.

Like many creative endeavours in music, audio samples might feature a blended timbre. Some samples exhibit vintage and modern sonic characteristics at the same time (Figure 1.12). For instance, a modern 808 kick might contain a high amount of saturation distortion in the low-middle (i.e., 200–500Hz) and middle frequencies (i.e., 500–1kHz). Or, alternatively, users might select an otherwise modern-sounding detuned vocal sample which happens to lack high frequency energy above 12 kHz. Some popular sample libraries which feature a hybrid vintage and modern sound include *Bedroom Beats and Lo-fi* (Capsun Audio), *Sonny Digital Drumkit*

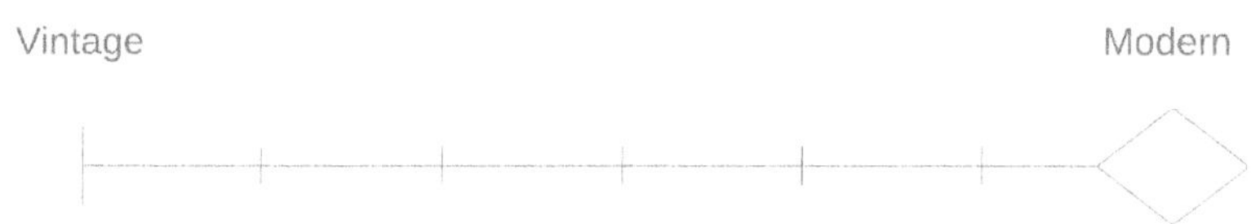

Figure 1.11 The right side of the spectrum: modern samples.

Figure 1.12 The middle of the spectrum: vintage and modern samples.

(Splice), *420: The Smoker's Kit* (Capsun Audio), *Organic Future Hip Hop* (Loopmasters), and *Ducko McFli Drum Essentials* (Splice).

Audio Example 1.10

In Audio Example 1.10, I demonstrate a number of sounds which demonstrate both vintage and modern sensibilities. Many hip-hop producers are fascinated with combinations of old and new sounds, and CBMP make this type of blending very easy to accomplish.

In addition to possessing a modern or vintage sound characteristic, samples may also be either timbrally (i) dense or (ii) sparse. A timbrally dense sample contains a wide spectral distribution, whereas a timbrally sparse sample features a more restricted range of frequencies. Quite often, if a sample contains multiple instruments or sounds, it tends to also possess a denser timbral quality. If fewer sounds are present, the sample tends to exhibit a sparse characteristic.

Samples which contain single instrument sounds such as a hi hats, a lead vocal, snare drum, or a melody played by a Wurlitzer (i.e., right hand only) are all examples of timbrally sparse audio segments. Samples which contain orchestras, bands and ensembles, or multi-octave distorted dubstep-style synthesizers, on the other hand, are each examples of timbrally dense sounds. Between these two extremes lay samples containing drum kits, layered percussion performances, partial groups and ensembles, very busy keyboard passages (over multiple octaves), and other similar sounds (Figure 1.13).

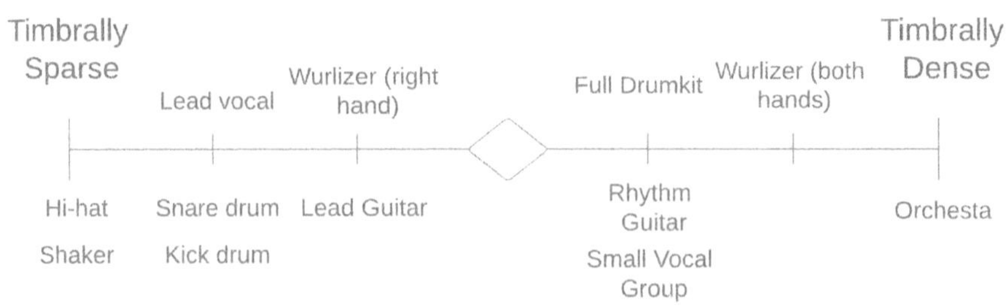

Figure 1.13 The approximate timbral density of some of the sounds mentioned in the above paragraph. Readers should note that exceptions to this classification system exist.

In the same way that bands and ensembles play arrangements which exploit the unique timbral strengths of each instrument in the group, the timbral qualities of an audio sample also determine how it will fit into a mix. For instance, if a producer is searching for some type of harmonic and melodic content to pair with drums and bass, s/he might seek out one or more samples which contain a keyboard and guitar. As heard within most rock bands, for example, keyboards and guitars primarily occupy a different frequency range than drums and bass. On their own, guitars, vocals, drums, and bass are each timbrally sparse enough to avoid masking one another. Together, the sounds from these instruments combine to form a single timbrally dense unit.

1.2.4 Categorizing Samples: Instrument Families as Conceptual Groups in Sample Arrangement

Having elaborated on CBMP arrangement practices, types of samples, and broad timbral categories in hip-hop, I now turn my attention to a more granular form of categorization. In sample arrangement, timbral characterizations such as vintage, modern, dense, and sparse describe a sound's frequency distribution, dynamic characteristics, and distortion type. Using these categories to discuss timbre in hip-hop CBMP has two benefits. On one hand, producers consider these dimensions either consciously or unconsciously during the beat-making process. As a result, for those interested in learning more about beat-making— whether for research or creative purposes—it can be useful to understand how a producer mentally categorizes these musical materials. This way of thinking provides insight into how producers hear records. For researchers who professionally analyze records, the adoption of this model can enable more accurate discussions of sample-based materials in hip-hop.

An additional way of conceptualizing sounds in CBMP requires one to consider the sound's general instrument *family*. For instance, a funky bass line played on a Fender P-bass, the low octave notes (i.e., the left hand) in a keyboard performance, and an 808-style bass can all be considered to belong within the *bass* family. Organs, Wurlitzers, Fender Rhodes, pianos, mellotrons, and Moog synthesizers are examples of *keys*.

CBMP libraries facilitate this type of organization by tagging samples according to the types of instrument sounds they contain. Users who intend to make music from these sample libraries will inevitably come into contact with this type of categorization. Splice, for instance, divides its entire library into the following categories: drums, percussion, vocals, synths, brass and woodwinds, guitars, keys, FX (i.e., effects), basses, and strings.

This type of instrumental grouping exists on a purely conceptual basis. Samples are not instruments—they are segments of audio. However, with this method of classification, producers can locate samples using readily understood categories.

Since many of the instrument categories listed above overlap, a sound might belong to multiple families at once. For example, if a label uploads a sample of a synthesized snare playing a bossa nova beat, CBMP services could simultaneously categorize it as a (i) percussion sample, (ii) drum sample, (iii) synth sample, and (iv) an FX sample. Users do not see this as a redundant feature since percussion, drums, synths, and FX are all different categories of *instruments*. Within a project, the sample could function as any one of these instruments, depending on the goals of the producer. In fact, CBMP services partly thrive because of this type of organizational flexibility, and services such as Splice, Loopcloud, and Noiiz heavily advertise this feature.

Instrumental categorization alone is not sufficient for modern music producers. As discussed earlier in this book, CBMP services offer a host of additional search filters. These filters allow users to hone in the search results they receive when using a CBMP library according to timbre, genre, and the sound's origination (i.e., acoustic, electric, synthesizer, field recording, etc.). When a sound is uploaded to a CBMP library, it receives a number of additional descriptive tags. These tags typically contain timbral (i.e., spectral) and aesthetic (i.e., genre-based) descriptors.

In a general sense, this type of search functionality reflects how producers think about sounds in a project. If someone requires a snare drum, s/he will search for "snare drum." S/he will also consider the timbre of the snare, which may be more vintage or modern (or terms which are interchangeable with these), and s/he may also consider the origin of the sound. For instance, a snare drum sample which came from a drum kit will sound quite different than a synthesizer snare which is comprised of white noise.

Since their arrival in 2013, CBMP services have been so successful, in part, because they facilitate music production in ways which music producers readily accept. By providing the ability to search and filter according to (i) instrument type, (ii) instrument timbre, (iii) instrument genre, and (iv) method of origin, these services streamline the sample-selection process for music producers. What was once an onerous *crate-digging* process—where producers would sift through volumes of vinyl records looking for the perfect sound—has now become a much simpler routine. In the next chapter, I elaborate on how producers use samples in order to create music.

1.3 SYNTHESIZERS

In addition to samples, hip-hop producers use synthesis to create and manipulate sounds. Perhaps the most famous synthesizer in hip-hop is the Roland TR-808, which found common use in 1980s and 1990s hip-hop scenes, as it still does today.[4] The 808 uses an internal synthesizer to provide a number of synthesized percussion sounds, which could also be programmed into sophisticated musical sequences.

Synthesizers are instruments that generate sound using devices known as oscillators. In the world of analog synthesis, these oscillators convert DC current from a power supply into an AC signal. Sonically, they produce basic sound waves such as sine, sawtooth, square, and triangle waves. Users can modify the sounds these oscillators create using various (i) amplifiers, (ii) filters, and (iii) modulation techniques. In a conceptual sense, digital synthesizers also work in a similar way. However, digital sine, sawtooth, square, and triangle wave oscillators rely on rapid mathematical calculation rather than the electronic components found in analog synths. Some examples of ubiquitous VST softwares which work this way include FM8 and Sylenth, for instance.

Audio Example 1.11

In Audio Example 1.11, using NI's Massive VST, I demonstrate the sound of 3 different oscillators. Listeners will hear an arpeggio sequence featuring one oscillator (Square-Saw 1), and the same sequence is repeated with the addition of a second oscillator (Polysaw 1) and a third oscillator (Woody). I have turned the amplitude up to *full* on each oscillator. Listeners should note how the timbre becomes more sophisticated as more oscillators are blended together.

In synthesis, sounds from oscillators are further modified by applying amplifiers, envelopes, filters, and strategies for parameter modulation. Of all these devices, amplifiers perform the simplest duty. Once an oscillator generates a sound signal, amplifiers simply boost signal amplitude.

Envelopes determine a signal's amplitude characteristics by controlling output levels as they unfold throughout time. The four parameters which allow synthesists to shape amplitude in this way are known as (i) attack, (ii) decay, (iii) sustain, and (iv) release. Each of these settings prescribe the signal's amplitude behaviour at distinct points in time, from the beginning to the end of its sounding life. When an envelope is applied to a sound generation device, like an oscillator, the attack and decay parameters determine the sound's initial onset amplitude characteristics. A long attack time causes the amplitude to increase slowly over time, whereas a short attack time does the opposite. Once this initial attack phase has ended, the envelope causes the signal's amplitude to decay according to the settings chosen by the user. Readers should note that in Massive (the synthesizer used for these demonstrations), the manufacturers have chosen to combine decay and sustain settings by adding a feature called "level" on both the attack and decay parameter settings (Figures 1.14–1.21).

Figure 1.14 An ADSR filter in Massive with the attack set at 0% (i.e., fast).

Figure 1.15 An ADSR filter in Massive with the attack set at approximately 25% (i.e., medium-fast).

Figure 1.16 An ADSR filter in Massive with the attack set at approximately 50% (i.e., medium).

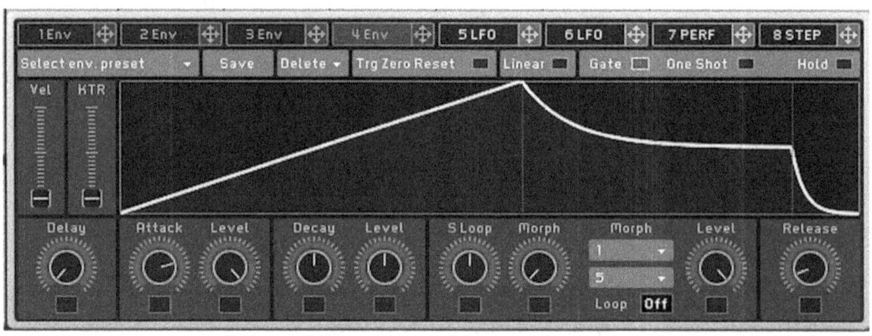

Figure 1.17 An ADSR filter in Massive with the attack set at approximately 75% (i.e., medium-long).

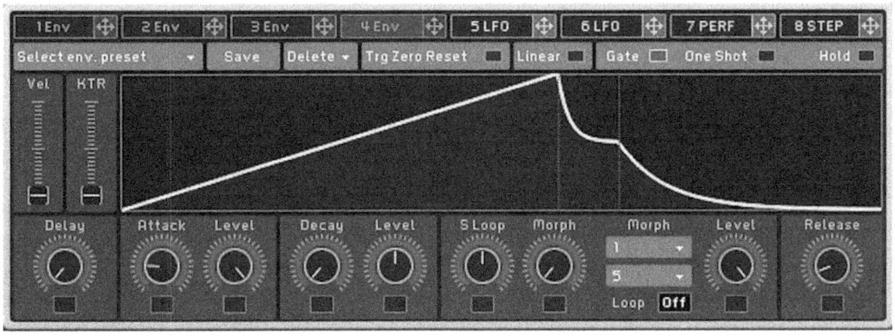

Figure 1.18 An ADSR filter in Massive with the decay set at 0% (i.e., fast).

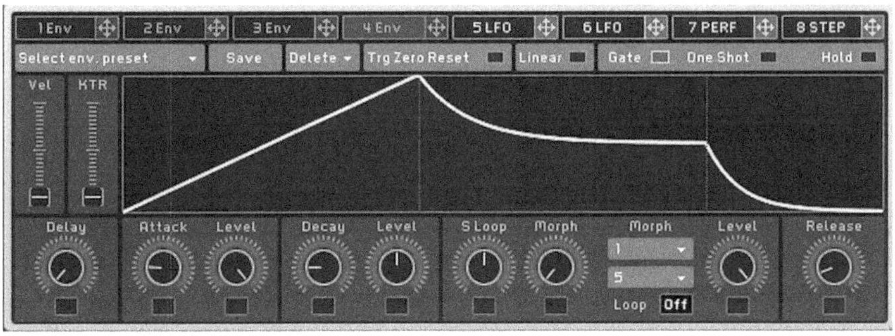

Figure 1.19 An ADSR filter in Massive with the decay set at 25% (i.e., medium-fast).

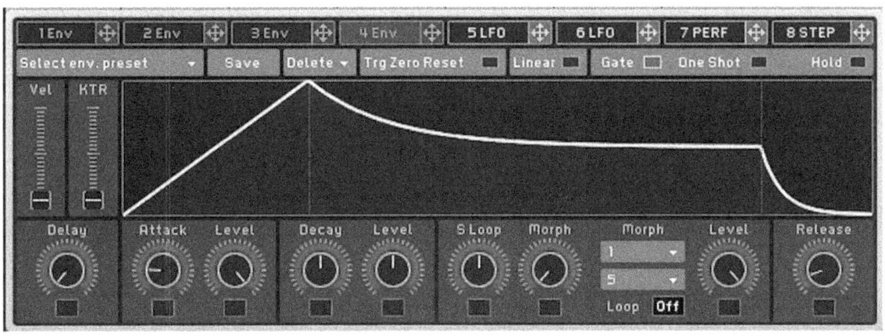

Figure 1.20 An ADSR filter in Massive with the decay set at 50% (i.e., medium).

Figure 1.21 An ADSR filter in Massive with the decay set at 75% (i.e., medium-long).

Audio Example 1.12

In Audio Example 1.12, I take the same patch I created using three oscillators in Audio Example 1.11, and then I apply a single envelope filter to all them. In the first part of this audio example, I experiment with different attack times (while decay, sustain, and release remain the same). For the first arpeggio sequence, the envelope's attack setting begins at 0. When the sequence repeats, I increase the attack to 25%. From this point, the attack increases to 50% and 75% for the following two repetitions of the arpeggio sequence.

For the second half of this audio example, I keep the attack set to 25%, and then I experiment with different decay settings. The first-time listeners hear the arpeggio sequence, and the decay is set to 0%. Each respective repetition then increases to 25%, 50%, and then 75%. More experienced readers should note that, in Massive, decay and sustain settings are effectively blended together through the provision of *level* settings in connection with each parameter (hence its absence from this demonstration).

From this point, the envelope enters its sustain phase. This parameter determines the amplitude level of the sound when a note is held for a period of time. If someone holds the note C4 (i.e., 261.6 Hz) on a synthesizer keyboard for 5 seconds, the sustain setting would determine the sound's amplitude for almost this entire time, except for the fractions of a second it takes for the envelope to go through its initial attack and decay phases. Predictably, an envelope's release setting determines the sound's amplitude behaviour once the user ceases input. Practically speaking, this normally refers to the moment when the person playing the synth removes a finger from a key on the keyboard (Figures 1.22–1.25).

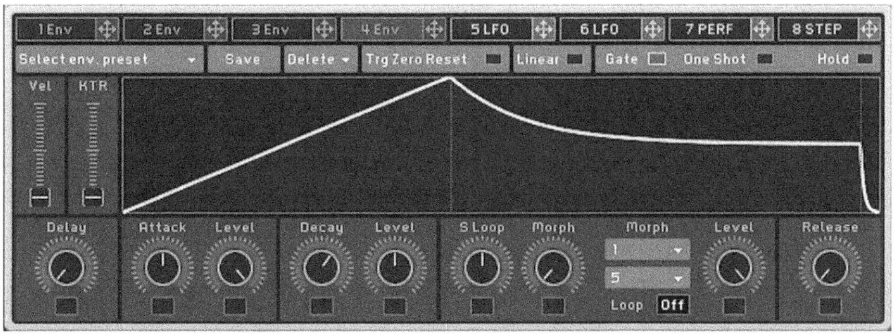

Figure 1.22 An ADSR filter in Massive with the release set at 0% (i.e., fast).

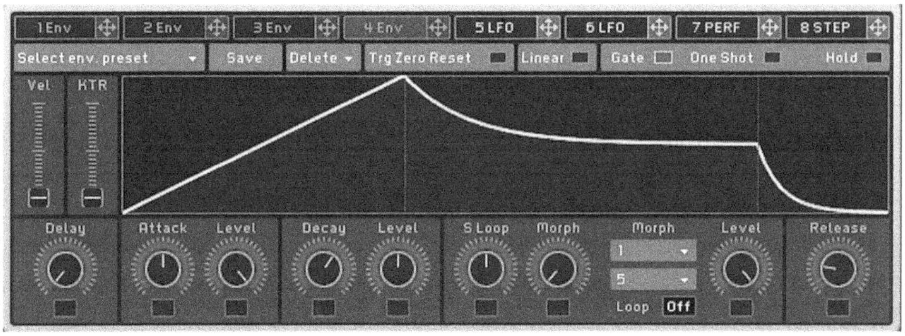

Figure 1.23 An ADSR filter in Massive with the release set at 25% (i.e., medium-fast).

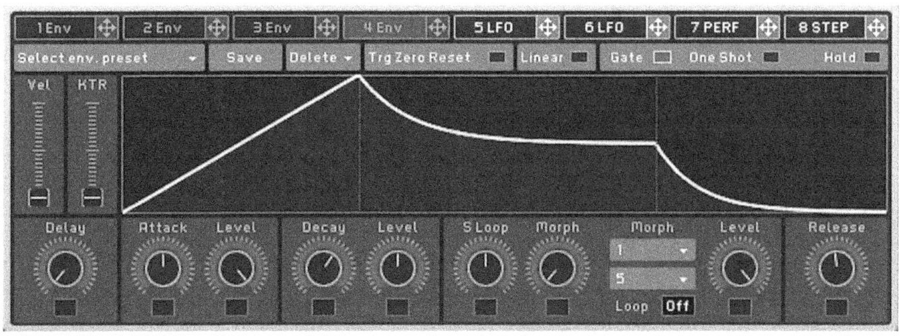

Figure 1.24 An ADSR filter in Massive with the release set at 50% (i.e., medium).

Figure 1.25 An ADSR filter in Massive with the release set at 75% (i.e., medium-long).

Audio Example 1.13

In Audio Example 1.13, I take the same patch I created using three oscillators in Audio Example 1.11, and then I apply a single envelope filter to all them. In this audio example, I experiment with various release times. The envelope's attack and decay settings remain near 50% for each example. For the first arpeggio sequence, the envelope's release setting begins at 0. When the sequence repeats, I increase the release to 25%. From this point, the release increases to 50% and 75% for the following two repetitions of the arpeggio sequence.

More experienced readers should note that, in Massive, decay and sustain settings are effectively blended together through the provision of *level* settings in connection with each parameter (hence its absence from this demonstration).

In Figures 1.22–1.25, I illustrate various sustain settings in Massive. In this particular software plugin, sustain and release parameters are blended together.

Envelopes (or envelope filters) shape a sound's dynamic (i.e., amplitude) characteristics. The term *filter* describes synth components which modify their harmonic or timbral output. Devices from this family include low and high pass filters, band pass filters, band reject filters, and others.

Perhaps the most engaging feature of synthesizers for producers is that these devices allow users to connect the various components, like oscillators, amplifiers, envelopes, and filters, in order to create sounds. In fact, many these circuit pieces can control the sound-shaping parameters of other items within the synthesizer circuit. For instance, one might cause a filter to work in accordance with the settings of an external ADSR envelope. The name for this technique is called *modulation*, which occurs in cases wherein a signal is shaped by the inputs of some outside source, or coefficient.

Modulation allows producers to alter the timbral properties of these instruments in novel and dynamic ways. If an ADSR envelope is assigned to a low pass filter's (LPF) cutoff frequency, then the LPF will remove high frequencies in accordance with the values assigned to the ADSR envelope. In the example below, I use the same synthesizer preset as the other audio examples in this section, and I add an envelope to a low pass filter. In the first half of this audio example, I play the sound with no additional modulation. After this occurs, I use a medium attack time on the low pass filter. Listeners will notice that each time a note plays, the timbre contains more low frequency energy before slowly regaining its lost high frequency energy. In the second half of this audio example, I increase the attack time on the envelope which controls the LPF. Each time a note plays, listeners will notice that it takes significantly longer for the device to regain high frequency energy than in the first half of this audio demonstration.

Audio Example 1.14

In Audio Example 1.14, I take the same patch I created using three oscillators in Audio Example 1.11, and then I apply an LPF to all of the oscillators. The cutoff frequency of the LPF is modulated by an envelope generator. By changing attack and release times on this envelope, the LPF responds in kind. At first, listeners will hear an arpeggio sequence with no modulation applied. After this occurs, I apply a fast-attack envelope to the LPF. Listeners will hear the same arpeggio sequence with this additional modulation. Once this plays, the sequence repeats with a slower-attack envelope applied to the LPF.

As heard in the demonstration, high frequencies are added back in to the original signal either slowly or quickly depending on the speed of the attack setting on the envelope.

The demonstration above shows only one simplistic application of modulation. With modern digital synths like Serum and Massive, one can modulate synthesizer parameters in endless ways. This allows producers to explore new sounds which cannot originate from any sort of acoustic phenomenon.

Since synthesizers offer so many sonic possibilities, CBMP libraries provide producers with large databases of synthesizer presets. Like with samples, these libraries organize presets according to the following instrument categories: drums, percussion, vocals, synths, brass and woodwinds, guitars, keys, FX, basses, and strings. This type of instrumental grouping exists on a purely conceptual basis. And, indeed, synthesizers are not guitars, basses, or percussion—they are sound-creating devices which can sound similar to these things if someone programs them to behave in such a way. Although ontologically inaccurate, this system of organization greatly aids producers in searching for presets with specific timbral qualities.

1.3.1 Synthesizer Types

A number of sound synthesis methods exist, and these different approaches are categorized according to the type(s) of waveform (or waveforms) which the oscillators generate, as well as the ways which the synthesizer handles the audio signals generated by each oscillator. In today's digital studios, producers typically use FM, subtractive, additive, and wavetable synthesizers.

FM synthesis relies on the combination of two signals: a *carrier* signal and a *modulator* signal. The carrier signal provides the fundamental frequency, as well as a portion of the sound's harmonic characteristics. The modulator signal introduces more harmonic sophistication and, thus, alters the timbre of the oscillators. The Yamaha DX7 is an example of an FM synthesizer which was quite common in the 1980s through the 1990s. Today, many producers use a popular digital plugin known as FM8 or Logic Pro's ESX24 (Figure 1.26).

Figure 1.26 Logic ESX24.

Audio Example 1.15

In Audio Example 1.15, I demonstrate a number of sounds created via an FM synthesizer. Listeners will hear a low computerized string sound followed by brass, bells, koto, and a harmonically rich pad.

Predictably, additive and subtractive synthesis methods work using opposite techniques. One method involves adding numerous sine waves together in order to generate sounds, and the other involves subtracting harmonics from a signal which has rich overtone characteristics. Additive synthesizers are less relevant in hip-hop production circles today; however, one well-known example is Loom by AIR Music Technology.

The core sonic principle behind subtractive synthesis, on the other hand, remains present in many of today's most popular synthesizers. Subtractive synthesizers generate sounds with dense harmonics before applying a series of filters which refine and focus the instrument's spectral output. The technique resembles whittling or clay making, wherein the removal of excess material causes the creation of new artistic objects.

Conceptually speaking, subtractive synthesis is relevant within the designs of popular wavetable synthesizers such as Spire, Serum, and Massive. CBMP services such as Noiiz and Splice, for instance, provide exhaustive lists of presets for these VSTs. These digital software synthesizers use oscillators which incorporate sound files known as wavetables. These files are actually short audio samples, and they contain sound waves which act as a set of instructions in order to determine the oscillator's sonic characteristics throughout time. The oscillator reads the wavetable file, which then effectively models the sonic output of the single-cycle waveform contained within this short recording. The advantage to this technique, unlike AM and FM synthesis, is that wavetable files already sound somewhat complete *before* the user adds any type of amplifier, filter, effect, or modulation strategy to the signal. As with other synthesizers, wavetable oscillators generate sounds which users modify by applying various amplifiers, filters, and modulation techniques (Figure 1.27).

1.3.2 Synths in Modern Hip-Hop

Today, the most common synthesized sounds in hip-hop (and closely related sub genres, such as trap) are 808-style sub basses, 808-style snare drums and claps, and 808-style kick drums, as well as plucks and pads (Shelvock 2017a: 176–177).[5] However, producers do not limit themselves to only using the 808 or sounds which emulate (and sometimes enhance) original 808 sounds. For instance, "plucks" and "pads" are also common, and producers create these sounds using a myriad of different devices. Equipment manufacturers and sound database designers use the term *pluck* to describe any synthesized sound with a staccato delivery. Pads, on the other hand, are sustained sounds that typically provide background harmony.

Figure 1.27 Wavetable oscillators from Serum.

Many sampled sounds are layered with synthesized sounds. For instance, producers commonly layer a saturated (and lo-fi-sounding) snare drum sample with a synthesized 808 snare drum. I have outlined some of these sounds, as heard on Travis Scott's "Butterfly Effect" (2017), in Table 1.2.

1.3.3 Hearing and Creating 808-Style Sounds

Originally, the internal synthesizer within the Roland 808 generated sound using an analog circuit; today, however, sounds like those described in the above chart are typically created using modern digital software synths. For instance, a Native Instruments' plugin known as Massive is a popular choice amongst producers for synthesizing drum sounds. CBMP services offer pre-programmed patches for popular synthesizers such as this, which users can incorporate within their own projects or modify further by simply adjusting the synthesizer parameters. Figure 1.28 demonstrates

Table 1.2 Travis Scott, “Butterfly Effect” (2017)

Time	*Sound*
00:00	808 sub bass
00:00	Synthesizer plucks*
00:13	Kick drum
00:14	808 snare drum (layered with clap)

*The producer applied a large amount of reverb and delay to these synthesizer plucks, and he also ensured that the notes sustained for a long duration. This also causes the main synthesizer melody (and counter melodies) to also sound like a pad.

Figure 1.28 A sub bass and kick drum combination patch in Native Instruments’ popular Massive software.

settings for a combination of an 808-style kick drum and sub bass called BS H8OHHH8.nmsv, which is available on Splice for Massive 1.3 and above.

Audio Example 1.16

In Audio Example 1.16, I use the above patch to create a trap bassline. Listeners will benefit from hearing this type of instrument in isolation, as these sounds are typically reinforced with various percussion instruments, as well as higher-register sounds, in released tracks.

Massive's internal software structure, as well as its control interface, utilizes a number of common synthesizer features which are mentioned earlier in this chapter, such as (i) oscillators, (ii) filters, and (iii) parameter modulation. In Massive, as with all synthesizers, the generation of sound begins with the oscillator. Most of the time, oscillators generate harmonically rich sounds including triangle, sawtooth, square, and pulse waves. However, oscillators may also generate more timbrally simplistic sounds, such as sine waves, which feature only a single harmonic (i.e., the fundamental pitch). In fact, sine wave-based instruments are crucial to hip-hop's sonic flavour, especially for kick drums and low 808-inspired sub basses.

In the above example, all three of Massive's oscillators are set to generate a sine wave. However, Massive uses wavetable synthesis, so these oscillators do not generate actual sine waves. Instead, the software uses audio data, similar to a sample, in order to model the sound of a simple sine wave. In the above patch, three oscillators synthesize hybrid sine wave sounds using wavetables. Two of these wavetables consist of sine waves blended with square waves (i.e., oscillators 1 and 2), and the third is blended with a triangle wave (oscillator 3). By using the "WT Position" knob on each oscillator, users can select a pure sine wave tone by moving the knob all the way to the left, or, alternatively, one can completely turn the knob to the right in order to generate a pure square or triangle wave sound. In Figure 1.28, oscillator 1 produces a pure sine tone, whereas oscillators 2 and 3 emit sine waves mixed with a square wave and triangle wave, respectively. From this point, users can use the "Intensity" knob to determine the sound's overall harmonic density. While these three oscillator's use Massive's "Spectrum" mode, the software also offers three separate "Bend" modes, which modulate the playback speed of the wavetable. By lowering and increasing the intensity knob, users can approximate the sounds of subtle FM synthesis modulation or pulse-width modulation.

NOTES

1. In this case, I use the term *real time* to describe performance aspects which are encoded when a producer begins recording. Actions which occur outside of real time can also impact these recordings, such as manually editing MIDI data or waveforms with a mouse and keyboard.
2. For example, please see Marvin Gaye vs Robin Thicke and Pharrell Williams: www.rollingstone.com/music/news/robin-thicke-and-pharrell-lose-blurred-lines-lawsuit-20150310 accessed September 27, 2017.
3. For an example of this technique, please see Mr. Carmack on the popular show *Rhythm Roulette*: www.youtube.com/watch?v=Hou9KevLeGU accessed September 25, 2017.
4. More often, producers now use samples from original 808 devices, or they synthesize their own similar sounds. At the same time, Roland TR-808s are now highly coveted vintage machines, and they are costly (approximately 4,000–5,000 USD).
5. Previously, Southern Rap, otherwise known as the Dirty South sound, often incorporated these sounds within their productions. Artists that rose to prominence in the early 2000s, such as TI, Ludacris, Rick Ross, Lil Wayne,

and many others, chose to use sampling less frequently than previous artists (even though sampling still occurs). In particular, the sounds of Paul Wall, Z-RO, and the Screwed Up Click were particularly influential on current trap and hip-hop emcees outside of the American south, including A$AP Rocky and Lil Uzi Vert.

2

Making Music Using CBMP Resources

2.1 MAKING MUSIC FROM SAMPLES

Of the production tasks discussed throughout this book, sampling, in particular, is perhaps the most commonly misunderstood activity. When the topic of sampling is mentioned in the news (and, especially within scholarly writing on the subject), it tends to only be discussed in terms of its legal and social implications. Since there is currently a scarcity of research regarding the ways that sampling is used in music production, in this section I demonstrate how music makers actually use sampling to make music. To do so, I cover a number of sample-based music-making techniques, and I unpack the musical implications for each one. In other words, I explain sampling as a musical technique, rather than as a social context or as a symbolic act of social resistance, as so many other authors have already done. With the recent proliferation of CBMP services such as Splice, Loopcloud, and Noiiz, sample-based music making is now more relevant than ever. However, misunderstandings about sample-based music making still remain prevalent amongst musicians and within peer-reviewed literature.

This situation arose, in part, because many authors who have written about sampling to date do not also make music using samples. As a result, the goal of this book, as well as others in the *Perspectives on Music Production* series, is to provide a practitioner account of the music production process, which explains, comments upon, and analyzes the act of creation in highly interdisciplinary way by drawing from science and technology studies, audio engineering, music production studies, human and computer interaction, and perhaps others.

As it stands, musicians and scholars tend to believe that sampling involves choosing one or two bars of pre-recorded music and then continuously looping this selection in order to create an instrumental track. But, as mentioned in this book's earlier chapters, this does not accurately represent the type of sampling practice that most producers use to create music. In order to address this misunderstanding, in this chapter I elucidate sample-based music making as a distinct species of musical practice.

To do so, I organize my explanation according the instrument-based classification mentioned in the previous chapter—namely, drums and percussion, vocals, synths, brass and woodwinds, guitars (and plucked instruments), keys, FX (i.e., effects), and strings. Of course, these categories are only conceptual in nature since those who create music from samples use pre-recorded audio, rather than instruments. Nonetheless, discussing audio samples in terms of the sounds they contain is a useful way to ground the current discussion in something familiar to readers. Readers should also note that, even though synthesis is also an important aspect of beat-making, I do not cover this topic to the same extent as I do with sampling. I avoid this topic because, for most readers, it will be readily apparent that one can simply download a synthesizer preset from a CBMP service and then use it to create a virtual performance of a flute, moog-style synth, violin, trumpet, or a host of other sounds. Instead, I focus on samples because it's still the dominant music-making technique in hip-hop and, until now, no peer-reviewed book has focused on the creative capacities and limitations of this creative activity.

In this chapter, I begin by explaining how producers use samples to create drum, percussion, and bass performances on records before moving on to vocals, synths, brass and woodwinds, guitars (and plucked instruments), keys, FX, basses, and strings. I proceed in this order because, in hip-hop, as with many other genres, drums and bass typically establish a project's overall groove (Shelvock 2017a: 178). Once the producer determines the project's groove features, s/he can begin to address melodic and harmonic components. In other words, the organization of this chapter reflects the typical workflow of a producer who works with samples.

After I cover melody and harmony, I also discuss another category of audio sample— those which contain *atmospheric* and *textural* sounds. Some of the samples in this category, such as *pads*, tend to have discernible pitch and chordal characteristics, whereas others contain non-pitched timbres, such as environmental noise or other sounds which are difficult to classify.

2.1.1 Sample Phrasing Charts and Audio Demonstrations

As mentioned in the previous chapter, I use the term sample phrasing to describe the ways that producers rhythmically arrange samples. Owing to some of the constraints of sampling technology (and, of course, recording technology), there are a number of recurring sample arrangement strategies that producers use to make beats. Throughout this chapter, I explain these approaches using *sample phrasing* and *sample arrangement* charts which reference some of the more common arrangement and timing strategies which producers use in popular records.

In addition to these charts, this chapter features numerous audio demonstrations. I created these examples in order to present various sampling techniques in isolation so that readers can train themselves to hear these techniques on other commercially available records. In addition, when discussing music-making practices *per se*, it is often futile to do so only using text. As Martin Mull stated, and as many musicians and composers have famously repeated since, writing about music can be like "dancing about architecture" (in Speranza 1979). By engaging with the audio materials I present in this chapter, it is my hope that readers can learn to hear and understand records as hip-hop beat producers hear and understand them.

In addition, this chapter will benefit readers interested in learning to use sampling techniques for their own musical goals. All of the samples used to create these demonstrations are available to readers via low-cost memberships to Splice, Loopcloud, or Noiiz. In some cases, readers may find similar samples for free on various sample production sites, such as Red Bull Academy, Cymatics.fm, and LANDR.com.

In the section below, I demonstrate some of the ways that producers use one-shots and continuous samples to create percussion parts, textures and atmospheres, melodies, harmonies, bass, vocal parts, and effects in order to create hip-hop beats. I do not provide an exhaustive list of how these activities are performed, nor do I offer secret knowledge of some type. Instead, I provide the opposite: what follows is an analysis of commonly used musical techniques which are crucial to a large portion of the world's music-making scene.

In fact, at the time of writing, one CBMP service known as Splice boasts over 2.5 million subscribers), and several other services exist (King 2019). This means that over 2.5 million music composers are currently *creating* music using CBMP sample libraries. And, of course, the music made with these tools is extremely popular for music fans since 1 out of every 3 music streams in the USA in 2018 was from hip-hop records (Resnikoff 2019).

2.2 DRUMS, PERCUSSION, BASS

In hip-hop, as with many other genres, drums and bass typically establish a project's overall groove. Yet, these instruments have a special relevance to the sound of hip-hop since most hip-hop records sonically highlight them. In fact, these two instruments may be the most important sounds in any commercial hip-hop track, apart from emceeing (Shelvock 2017a: 178). In this section, I explain how to create drums, percussion, and bass sounds using samples I obtained from CBMP sources.

2.2.1 Creating Drum and Percussion Tracks With One-Shots: Timbral Selection and Arrangement

Most one-shot samples are recordings of drums, such as kick, snare, and tom drums, as well as various types of cymbals, such as hi hats, crashes, and rides. There are also countless other percussion sounds available in a one-shot format, such as woodblocks, whistles, cracks, pops, footsteps, and water drops to name only a few. Producers create sequences from these short recordings, which form the rhythmic basis for hip-hop instrumentals. I provide examples of these sounds in the previous chapter.

For each type of percussion sound listed above, there are countless samples for producers to choose from. These samples may possess any conceivable timbral profile. Kick drums available on services such as Splice, Noiiz, and Loopcloud, for instance, can range from *big* and *boomy* 808 drums to *flat* and *warm* drums which resemble drumkits from the early 1970s. As I cover more thoroughly in Chapter 1, of the many existing timbral possibilities for drums, producers gravitate towards two dominant timbral *species* in hip-hop percussion. On one hand, producers may choose samples which are recorded or processed to sound more like *vintage* drum recordings. These samples often contain distortion, as well as a timbral profile which resembles the sound of drums being played back on a vinyl record or tape player. Alternatively, it is just as common for producers to use *cleaner*-sounding samples, which are either modern high quality digital recordings of drumkits or electronically generated sounds. I refer to this second timbral species as *modern* samples (Figure 2.1).

The audio examples below demonstrate drumbeats made from one-shot samples. For each, I provide (i) a photo of the corresponding MIDI track, (ii) an audio demonstration, and (iii) a frequency analysis of each example's audio data (Figures 2.2–2.7). The MIDI data will help the listener understand which samples are being triggered throughout the example, and the frequency analysis will instruct the reader regarding the overall spectral configuration of these drums.

Audio Example 2.1

Audio Example 2.1 demonstrates a beat made from one-shot samples which feature vintage timbral characteristics. This passage mimics the style of the hip-hop producer J Dilla, whose style is particularly influential on today's "lo-fi" and "chill" hip-hop scenes (Watson 2019). The timing of the kick drum sounds as though it occurs milliseconds *behind* the beat, and the snare drum sounds as though it occurs milliseconds *ahead* of the beat. In addition, the hi hats play *swung* eighth note pairs, but the timing is not the standard swing rhythm (i.e., a *triplet* feel, wherein the first two notes are tied together). This groove is known to hip-hop drummers and percussionists, such as Questlove, as the *drag* beat.

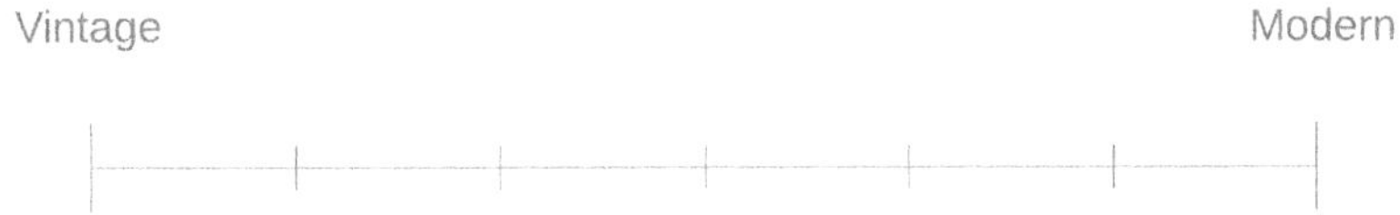

Figure 2.1 Reproduction of the vintage/modern continuum from Chapter 1.

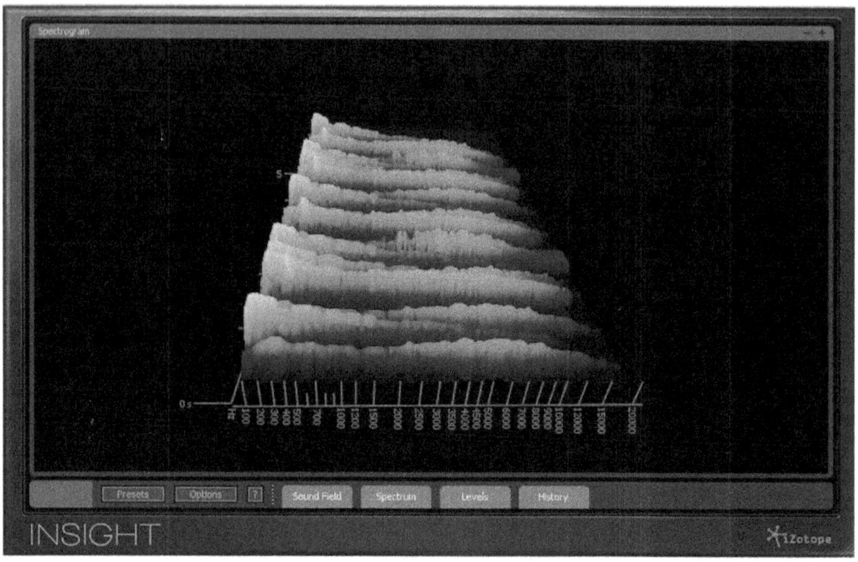

Figure 2.2a Spectral analysis of Audio Example 2.1 over several seconds.

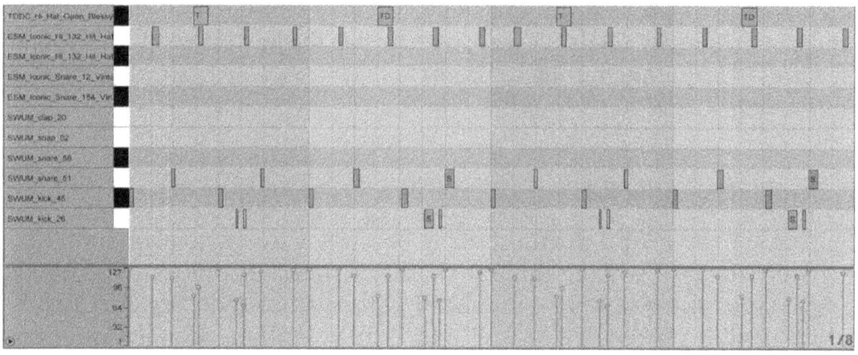

Figure 2.2b the corresponding MIDI for the drum performance in Audio Example 2.1, created from one-shot samples.

Audio Example 2.2

Audio Example 2.2 demonstrates a beat made from one-shot samples which feature vintage timbral characteristics. The kick drum features a clear pitch (D♯/E♭), which, when combined with its distorted features, causes it to stand out in the mix. There are also two snare drums in this passage. The first snare drum has a richer timbral characteristic, featuring an abundance of spectral energy around 300–600Hz. This drum occurs on beats 2 and 4. The second snare drum provides a more energetic rhythmic figure which mostly occurs on the so-called *upbeat*. This snare drum resembles a vintage 808-style snare drum, with its frequency make-up focused around the mids and upper mids.

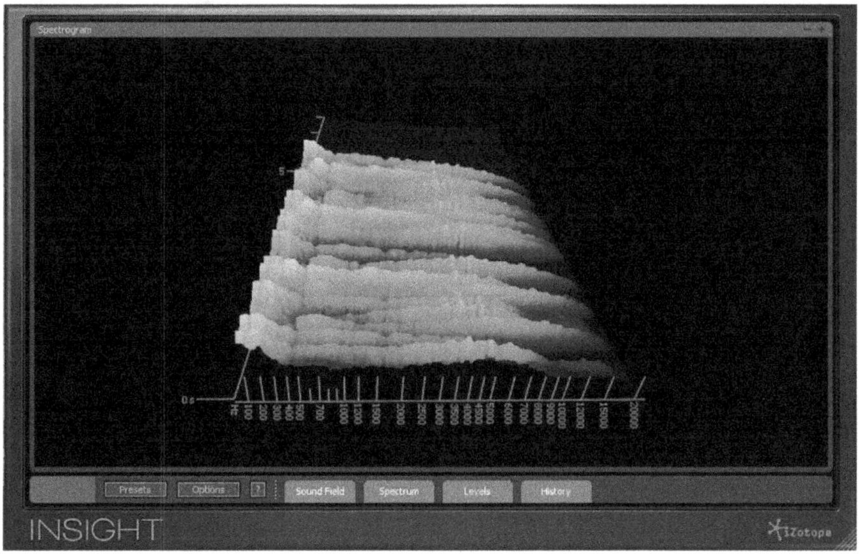

Figure 2.3a A spectral analysis of Audio Example 2.2 over several seconds.

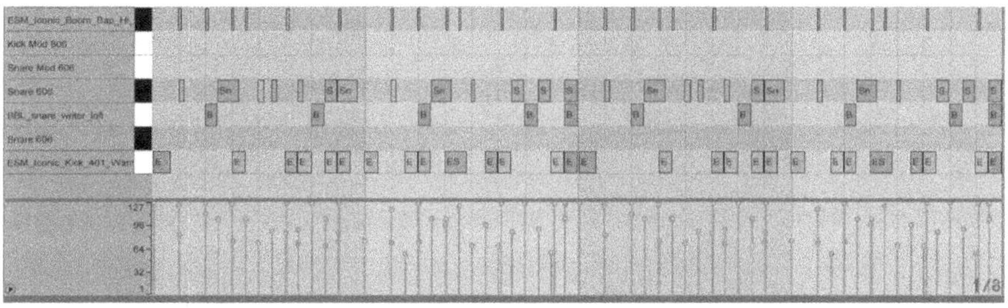

Figure 2.3b The corresponding MIDI for the drum performance in Audio Example 2.2, created from one-shot samples.

Audio Example 2.3

Audio Example 2.3 demonstrates a beat made from one-shot samples which feature modern timbral characteristics. More specifically, this beat belongs to the so-called *trap* subgenre of hip-hop. In general, producers in this genre prefer louder (i.e., compressed or with a higher average amplitude) samples with a wide frequency distribution—from the lowest audible sub bass frequencies to 20 kHz. The drum and bass samples sonically resemble samples produced by a Roland 808. However, these particular samples are both louder and more distorted than the original 808 sounds.

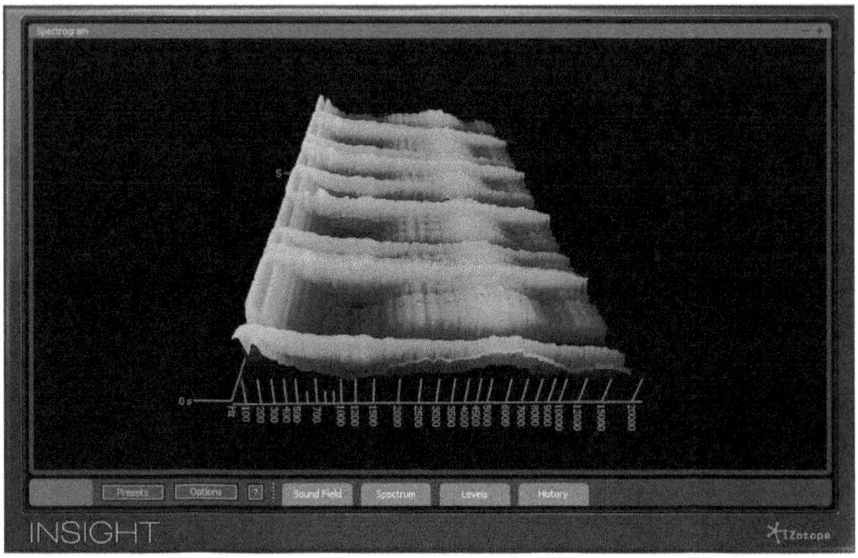

Figure 2.4a A spectral analysis of Audio Example 2.3 over several seconds.

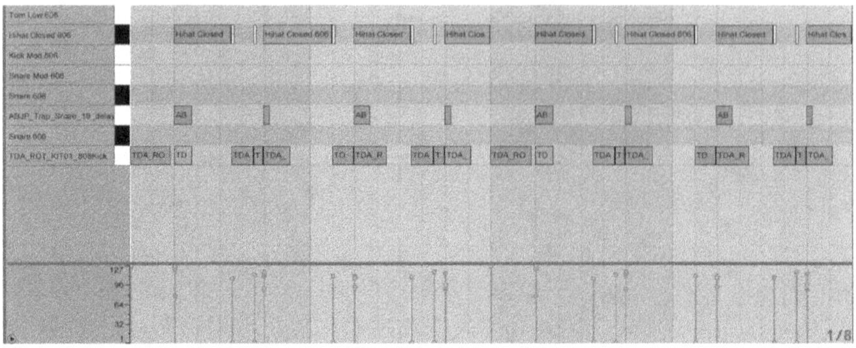

Figure 2.4b The corresponding MIDI for the kick and snare in Audio Example 2.3, created from one-shot samples.

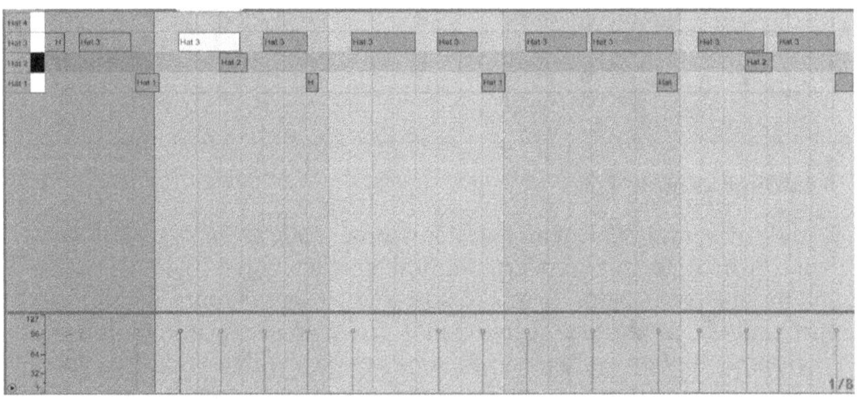

Figure 2.4c The corresponding MIDI for the hi hat performance in Audio Example 2.3, created by loading one-shot samples into a sampler vst. This vst causes the sample to repeat at predetermined rhythmic values. In this case, those values are 64th notes (hi hat 1, bottom), 32nd notes (hit hat 2, middle), and hi hat 3 (hi hat 3, top).

Audio Example 2.4

As with 2.3, Audio Example 2.4 demonstrates a trap beat made from one-shot samples which feature modern timbral characteristics. In this example, the sub bass also resembles the sound of a Roland 808 sub bass; however, this sample has been distorted to an extreme level. This causes the bass to stand out in the mix by highlighting (and perhaps even creating) its upper harmonic frequencies. Moreover, when producers distort the bass this way, it also becomes far more audible on a variety of common listening platforms, such as laptop speakers, phone speakers, and small earbud headphones.

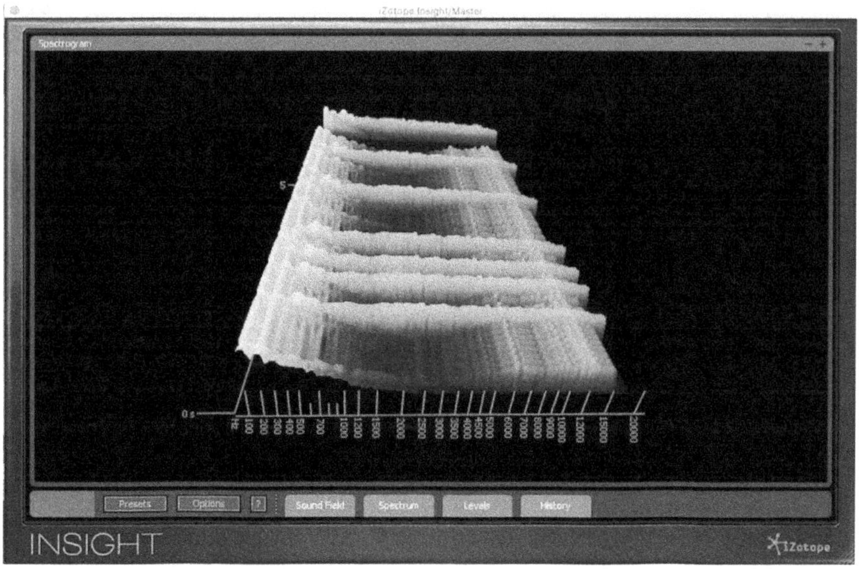

Figure 2.5a A spectral analysis of Audio Example 2.4 over several seconds.

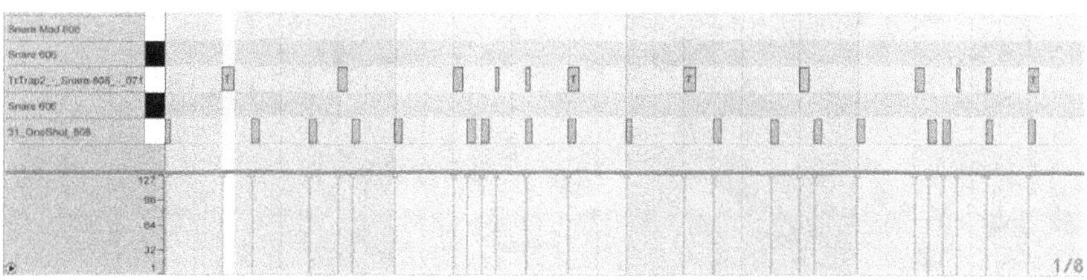

Figure 2.5b The corresponding MIDI for the kick and snare in Audio Example 2.4, created from one-shot samples.

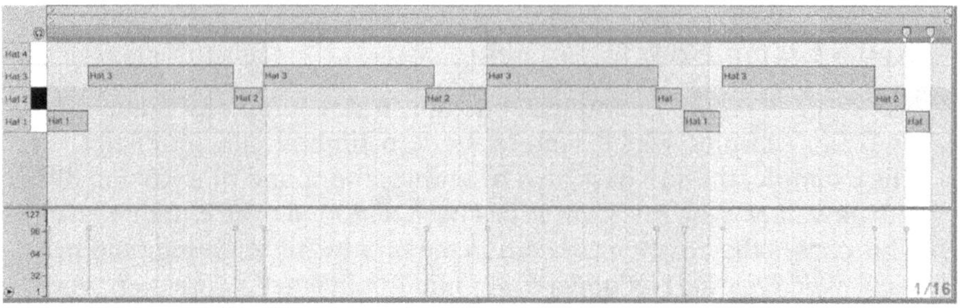

Figure 2.5c The corresponding MIDI for the hi hat performance in Audio Example 2.4, created by loading one-shot samples into a sampler vst. This vst causes the sample to repeat at predetermined rhythmic values. In this case, those values are 64th notes (hi hat 1, bottom), 32nd notes (hit hat 2, middle), and hi hat 3 (hi hat 3, top).

Audio Example 2.5

Audio Example 2.5 demonstrates a beat made from one-shot samples which simultaneously feature modern and a vintage timbral characteristics. These drums feature a wide frequency distribution; however, the kick and snare drum samples aren't as *clean* (i.e., a lack of distortion) as heard in Audio Example 2.4. These two samples, as well as the block percussion sample, exhibit a more vintage timbral characteristic, whereas the bass exhibits a more modern one.

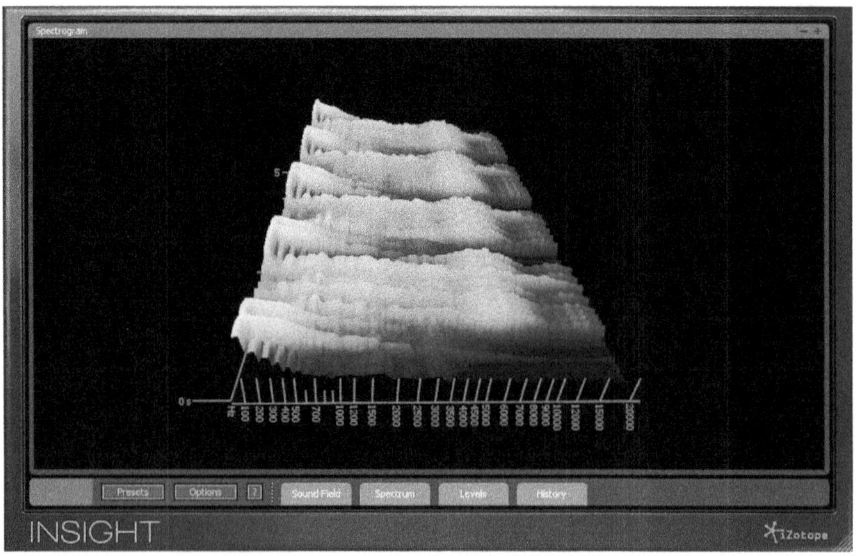

Figure 2.6a A spectral analysis of Audio Example 2.5 over several seconds.

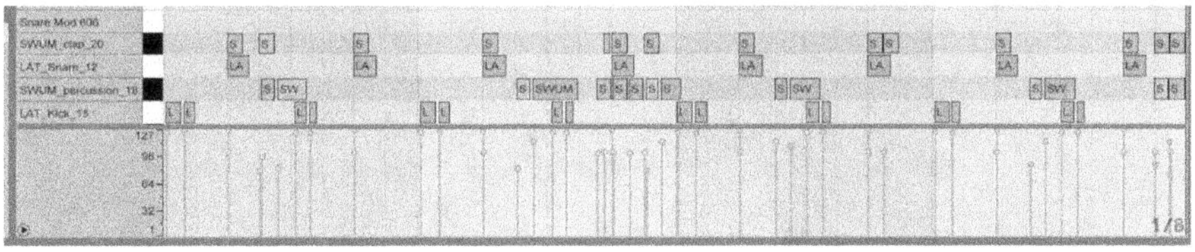

Figure 2.6b The corresponding MIDI for the kick and snare in Audio Example 2.5, created from one-shot samples.

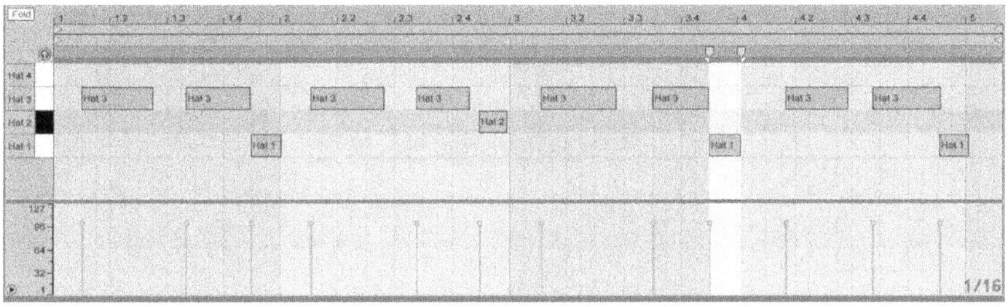

Figure 2.6c The corresponding MIDI for the hi hat performance in Audio Example 2.5, created by loading one-shot samples into a sampler vst. This vst causes the sample to repeat at predetermined rhythmic values. In this case, those values are 64th notes (hi hat 1, bottom), 32nd notes (hit hat 2, middle), and hi hat 3 (hi hat 3, top).

Audio Example 2.6

Audio Example 2.6 demonstrates a beat made from one-shot samples which simultaneously feature modern and a vintage timbral characteristics. The snare drum, cymbals, and block percussion samples feature a vintage characteristic, whereas the kick drum and the bass feature a modern characteristic. This particular blend of vintage and modern percussion samples is becoming increasingly common in modern hip-hop, as heard throughout J Cole's 2018 album *KOD*.

2.2.2 Rhythmic Congruence in Sampling: Using Continuous Samples for Additional Percussion

Although producers typically use one-shots to form the basis of hip-hop drum grooves, they may also include continuous samples. Most often, these continuous samples feature performances by various auxiliary percussion

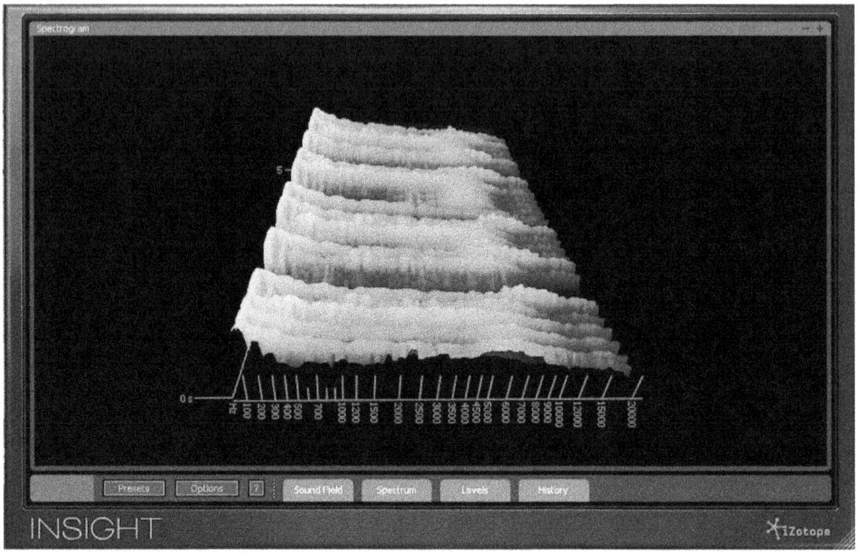

Figure 2.7a A spectral analysis of Audio Example 2.6 over several seconds.

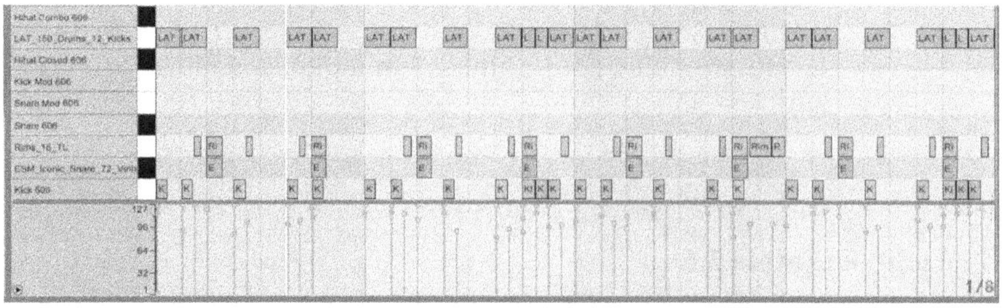

Figure 2.7b The corresponding MIDI for the kick and snare in Audio Example 2.6, created from one-shot samples.

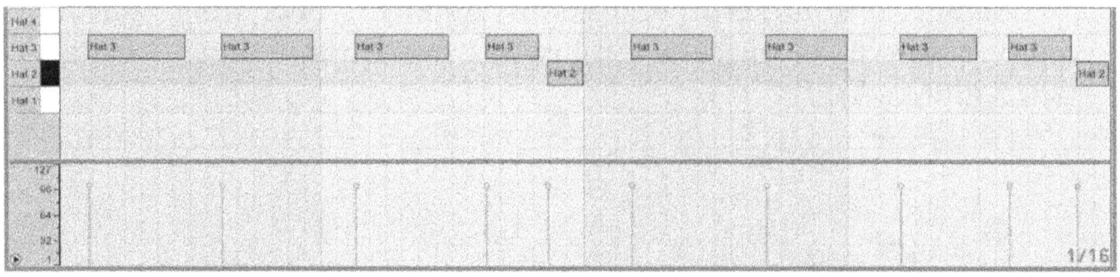

Figure 2.7c The corresponding MIDI for the hi hat performance in Audio Example 2.6, which was created by loading one-shot samples into a sampler vst. This vst causes the sample to repeat at predetermined rhythmic values. In this case, those values are 64th notes (hi hat 1, bottom), 32nd notes (hit hat 2, middle), and hi hat 3 (hi hat 3, top).

instruments, such as congas, shakers, claps, woodblocks, synthesized percussion sounds, and others. When producers use these sounds, they typically do so in order to add rhythmic sophistication into their instrumental tracks.

These continuous samples may contain recordings of percussion performances which are either rhythmically congruent or incongruent with the rest of the instrumentation. Rhythmically congruent sounds are sounds that, when compared to one another, neatly fit within the digital time grid. These sounds are also easily expressed using standard musical notation as whole notes, half notes, quarter notes, eighth notes, triplets, sixteenth notes, thirty-second notes, and other standard rhythmic values.

Audio Example 2.7

In Audio Example 2.7, I match a kick, snare, and bass groove I created using one-shots with a rhythmically congruent continuous sample. In this case, I mean that the sample's timing characteristics closely match the groove I created using one-shots.

At first, listeners will hear the groove I created using one-shots. After 11 seconds, the continuous sample begins. Listeners will notice that the rhythms match one another. At 24 seconds, the continuous sample disappears for a moment to allow listeners' ears to readjust. After four beats, the continuous sample begins again.

For readers who have access to CBMP services, this example can be re-created using the following samples: DSDS_120_Iyesa_Shaker4–1.aif; Shh_brk98_dillatime_stp.aif.

Samples which are rhythmically congruent with the other elements in a project require very little sonic adjustment, apart from perhaps a mixing treatment. Alternatively, continuous samples which are rhythmically incongruent with other sounds in a project may require some additional changes in order to blend with the rest of the arrangement. I demonstrate this type of situation in Audio Example 2.8.

Audio Example 2.8

In Audio Example 2.8, I place a continuous kick and snare sample with a rhythmically non-congruent continuous shaker sample. Listeners will notice that the sound of these two samples together is fairly chaotic but, at the same time, quite lively energetic.

If a producer wants to add a shaker loop which does not rhythmically align with the kick drum, as within Audio Example 2.8, there are four options available. Producers may (i) choose to preserve the non-congruent rhythms because they sound appealing, (ii) select a new sample, (iii) edit the original sample so that it becomes more rhythmically congruent with the rest of the project, or (iv) apply sidechain compression to reinforce the track's main groove elements. In the example below, I cause the track to become more rhythmically congruent with the kick and snare drum via editing:

Audio Example 2.9

In Audio Example 2.9, I replay the audio samples from Audio Example 2.8 once again. However, this time, the shaker sample has been edited using Ableton's groove pool feature. I used Ableton's "extract groove" procedure in order to record the rhythmic data of the kick and snare sample (quantified as transient detail over time). After this occurred, I simply applied the groove to the shaker sample using Ableton's groove pool. This caused the two samples to rhythmically align with one another.

A producer may use sidechain compression to cause the shaker's amplitude to reduce in conjunction with the timing of the drum groove. The procedural aspects of this technique are discussed more thoroughly in this book's chapter on mixing. However, I explain some of the rhythmic consequences of the technique in this section on arranging samples.

A compressor is an automatic gain controlling device. *Sidechain compression* is a technique which causes one track (i.e., signal A) to sound quieter in response to amplitude fluctuations in another track (i.e., signal B). If a project contains a kick, snare drum, and shaker, for instance, sidechain compression can cause the shaker's amplitude level to drop each time the kick drum occurs. As the shaker's amplitude levels are restored, it causes the audio *pump*. By causing the shaker's loudness levels to fall predictably each time the kick drum occurs, the shaker effectively takes on some of the drum's rhythmic features. And, at moments in the project where the kick drum does not occur, the shaker will continue to sound rhythmically incongruent with the rest of the arrangement. This allows the producer to gently reduce the level of rhythmic misalignment between these elements at crucial moments while also maintaining a rhythmically unrestricted sound at times.

Audio Example 2.10

In Audio Example 2.10, I pair kick and snare sample with two rhythmically non-congruent shaker samples. At first, listeners will hear that the shaker samples sound somewhat distracting and overbearing. At 19 seconds into the recording, I apply sidechain compression to these samples, which I have configured to respond to the kick drum. This causes the gain of the shaker samples to be attenuated each time the kick drum occurs. Listeners will notice that the shakers sound more congruent with the kick and snare drum in the second portion of the recording (i.e., after 19 seconds have passed). This attenuated gain is returned shortly after the kick drum sounds, creating a rhythmic *upstroke* (Hodgson 2011). Readers can re-create this example using the following samples:

Percussion Samples:
Zoid_fx_loop_glitch_02_109
Zoid_percussion_loop_poppin_132
Kick and Snare Samples:
ASD_Kick___Snare_Loop_95_Tech_01

2.2.3 Hip-Hop Micro Rhythm: Creating Intentional Rhythmic Incongruence

Many hip-hop records use instrumentals which drift from the *perfect* time grid which digital technology enables. In fact, according to hip-hop/R&B producer and singer Georgia Anne Muldrow, some beat makers prefer to work "off the grid," which means that they chose to avoid creating drum tracks which perfectly align with a DAW's digital time grid. As she explains (in Houghton 2017):

> Off the grid—that's the way I have my system set up. Unless it's something industrial I want to do. . . . I try to make sure I'm relaxed enough to be able to see time for what it is, you know? That's how you can get the kind of detail you want in your drum pattern, if you can stretch time in your own head. You know, you got the metronome going and you can hear it fast or slow, even though the time's the same. It's *you* bending around the time. It's a blessing and that part comes from the ancestors; they gifted me with that the ability to stretch around a phrase in all types of different ways, like *The Matrix* when he's stretching around the bullets. It's just a point of relaxing yourself.

Indeed, as Muldrow explains, some producers adopt a highly flexible rhythmic approach, even though a common criticism of digitally created music is that it features overly rigid rhythmic content. Elsewhere, researchers have verified this phenomenon in hip-hop and R&B music (Shelvock 2017a: 175). Research by Anne Danielsen (et al.) at the University of Oslo, for instance, confirms the existence of this micro rhythmic approach in hip-hop in empirical terms (2010).

This type of "off the grid," or rhythmically *free*, approach was popularized by J Dilla. A well-known innovator, Dilla had a major influence on how beat makers structure and present rhythm in their work. According to DJ HouseShoes, who worked with him in the 1990s (in Russonello 2013):

> He was one of the first cats that kind of broke down the rigidity and the rules and the boundaries of hip-hop. Hip-hop had a stiff, structured code to it, and that definitely got loosened up after his reign.

His production style helped shape the next generation of beat makers, many of whom later flourished within the LA beat scene, such as Flying Lotus and Thundercat. Today, in the same scene, Dilla-esque rhythms prevail in beats created by Tokimonsta, Knxwledge, and Ras G. In more global terms, a Dutch label with an international following known as Chillhop only releases music which features this type of micro rhythmic strategy.

Today, a number of "off the grid" approaches coexist in the modern hip-hop scene. However, the prototypical J Dilla drum sound involves the use of hi hats and snare drums which are slightly ahead of the digital time grid and kick drums which freely land behind the beat. In addition, classic Dilla-style hi hats and other drums, which play sequential eighth notes, are often *swung* to varying degrees (Shelvock 2017a: 175).

However, this type of swing does resemble the typical eighth note swing, wherein the first two notes of a triplet are simply tied together. Instead, the prototypical Dilla beat I discuss above is best experienced via listening. The following examples each use the same kick, snare, and hi hat pattern with slightly different micro rhythmic variations in order to demonstrate these rhythms. In these demonstrations, listeners will hear (i) the kick drum on beats 1 and 3, (ii) the snare drum on beats 2 and 4, and (iii) hi hats on every upbeat (i.e., unaccented eighth notes). However, the groove sounds quite different in each example as I move each of these sounds (i-iii) away from a perfect digital time grid. At first, I place the hi hat and snare drums slightly ahead of the beat in order to simulate the rhythmic styles popularized by J Dilla and Madlib. I then cause the hi hats and snare to occur behind the beat in order to simulate the techniques heard on newer sampled-based hip-hop songs such as "The Waters" by Mick Jenkins (2014).

Audio Example 2.11

Audio Example 2.11 demonstrates a J Dilla style drumbeat which I created by recording my performances on a MIDI controller. J Dilla famously followed a similar routine using the classic MPC3000. I have intentionally tried to exaggerate some of the features of the classic Dilla beat, wherein the kick drum sounds as though it falls behind the beat and the snare sounds as though it occurs ahead of the beat (Figure 2.8).

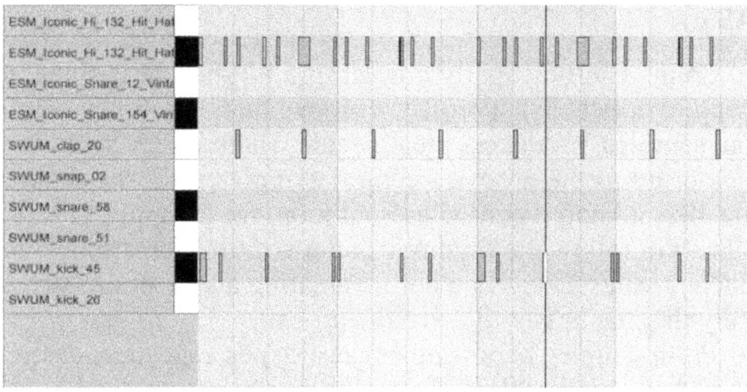

Figure 2.8 Screenshot of the MIDI data which corresponds to Audio Example 2.11.

Audio Example 2.12

Audio Example 2.12 demonstrates a J Dilla style drumbeat which I created by using the point-and-click method. A number of modern drummers, such as Questlove, have noticed that the J Dilla groove can be re-created using a polyrhythmic feel based on groupings of 7. So, in order to create this example, I created a hi hat figure which occurs 7 times within 4 beats (in 4/4 time). This means that every 4 beats, the septuplet groove effectively *resets* as the kick drum returns to beat 1. The snare drum lands on the 3rd and 7th hi hat within this rhythmic grouping, as pictured within Figure 2.9.

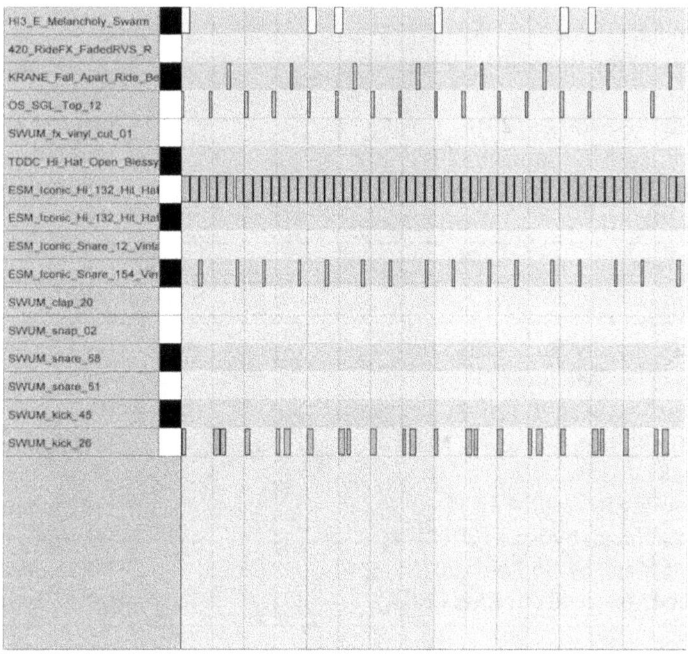

Figure 2.9 Screenshot of the MIDI data which corresponds to Audio Example 2.9.

2.3 BASS

Bass and drums can be equally crucial to a hip-hop beat since, together, both instruments establish a project's groove features. When creating bass parts from samples, producers must decide whether they will use the bass or the kick drum as the groove's central component. If the producer adds a bassline after having recorded a drum loop, for example, it is likely that the bassline will follow the rhythmic pattern of the kick drum. However, if the producer creates a bass line before creating a kick drum part, then it is more likely that the bass will drive the groove.

As with drum samples, bass samples can also possess a more *modern* or *vintage* timbre. And, predictably, different hip-hop subgenres tend to favour different approaches. Classic hip-hop tracks, and those which emulate the sound of these records, tend to favour a more vintage approach. It is worth noting that newer releases can still capture a classic hip-hop sound. For instance, the basslines heard on 2018 and 2019 Chillhop releases by Philanthrope, CYGN, and L'Indecies all make use of a more vintage bass sound (see Table 2.1).

Table 2.1 Bass Sounds With a *Vintage* Aesthetic

Vintage Bass	
Song/Artist	*Timing*
"Fu-Gee-La" by The Fugees (1996)	00:00 (throughout)
"Simon Says" by Pharoahe Monch (1999)	00:11, bass begins
"Hip Hop Hooray" by Naughty by Nature (1993)	00:09, bass begins
"Chief Rocka" by Lords of the Underground (1993)	00:00 (throughout)
"Numbers on the Boards" by Pusha T (2013)	00:05, bass begins
"Suede" by NxWorries (2016)	00:02, bass begins
"Old Beat" by Mr. Carmack (2011)	00:23, bass begins
"The Cut Off" (feat. kiLL edward) by J. Cole (2018)	00:03, bass begins
Modern Bass	
Song/Artist	*Timing*
"Vice City" by Jay Rock (2015)	00:17, bass begins
"Hallelujah" by Mouthe (2016)	00:32, bass begins
"The Most" by SO Y.T. (2018)	00:00 (throughout)
"Going Bad" (feat. Drake) by Meek Mill (2018)	00:06, bass begins
"WIN" by Jay Rock (2018)	00:15, bass begins
"The Flute Song" by Russ (2018)	00:14, bass begins
"Elephant Man" by So Loki (2018)	00:26, bass begins
"Connect 4" by the Cool Kids (2016)	00:11, bass begins

Audio Example 2.13

In Audio Example 2.13, I demonstrate a number of vintage bass samples. The first 19 seconds demonstrate vintage bass samples, whereas the remainder of the recording demonstrates modern bass samples. Readers who have access to CBMP services can search for the following files to find these samples:

Vintage Bass Samples:
TWIN_SHADOW_bass_guitar_loop_cola_118_Dmin
PRO_F#_OrganBass
AED_90_A_Afrobeat2_Bass_Guitar
PRO_D_HeliBass
AED_90_A_Afrobeat2_Bass_Guitar
Modern Bass Samples:
BBL_75_808_sub_trap_trees_Dm
BBL_75_808_sub_gliding_houses_A
FKI_glory_808_loop_135
OS_EHHM_125_Gmin_808_Line_24

Depending on whether the kick drum or the bass provides the focal point of a project's groove, producers may use different arrangement strategies. On more current hip-hop and trap records, for example, modern-sounding 808 basses typically follow the same rhythmic pattern as the kick drum. These bass sounds last for a long duration, whereas kick drums tend to be much shorter audio files. A single 808 one-shot sample, such as Standalone-Music_-_808_-_Spring_C1.wav (from the pack called *Boom!* By Standalone Music), can be assigned to a sample-triggering VST. From this point, producers can use MIDI to record a bass performance, and, to do so, they may either use the point-and-click method or a sample-triggering device.

Audio Example 2.14

On trap records, kick and bass samples often follow the same rhythmic pattern. In Audio Example 2.14, I demonstrate this approach. Figure 2.10 provides an illustration of this arrangement approach.

Producers can use a similar strategy with vintage samples, as long as the bass sample sustains over successive beats. However, in cases where one desires a more sophisticated bass sound, CBMP libraries offer a host of bass guitar performances. These libraries contain samples of nearly any conceivable music genre, and this provides producers

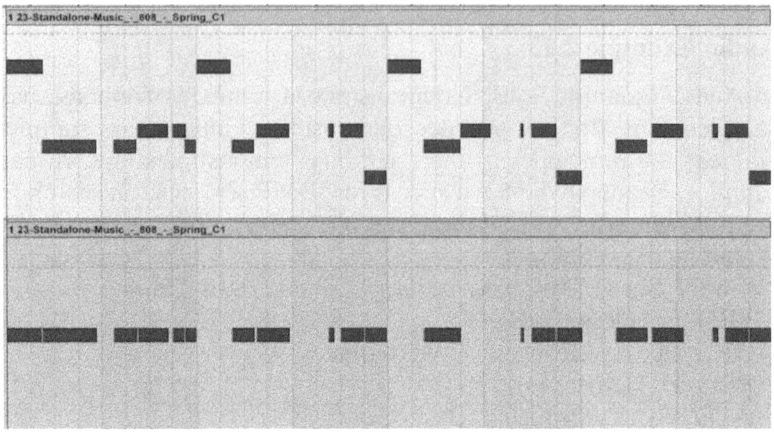

Figure 2.10 Kick and bass arrangement in trap music.

with a rich repository of musical performances. For instance, sample packs such as Bedroom Beats & Lofi Hip-Hop by Capsun Audio contain a plethora of recordings of jazzy stand-up bass performances, whereas Twin Shadow Siker Pack by Splice has a number of funky p-bass performances.

In terms of arrangement, the simplest way to make a *busier*, or timbrally *full*, sounding bass compliment the drums within a project is to begin with a simple snare and hi hat. From this point, one can import a bass performance, such as "DLTT_808_Sub_Bass_07_140_C.wav" (from Down Low's *Trap & Twerk* sample pack), or, perhaps, they might create a bassline themselves). Once these two tracks are in place, a producer can experiment with various kick drum patterns until s/he is satisfied with the results.

Audio Example 2.15

To create the groove in Audio Example 2.15, I began with a continuous snare sample (00:00–00:13). From this point, I imported a continuous hi hat sample (00:13–00:26). In this recording, the rhythm of the hi hat performance has a clear trap influence, as heard when it alternates between eighth note and sextuplet rhythms.

I add a continuous 808-style trap bass sample between 00:26–00:39 followed by a kick drum. The kick drum reinforces the rhythm of the bass, and it occasionally occurs on its own in order to add some rhythmic interest to the track. Readers who have a CBMP subscription can re-create this example using the following samples:

Hi hat:
Standalone-Music_-_Destiny_HiHat_Loop_13_150BPM
Snare:
Ct_drm140_anti_snr
Bass:
DLTT_808_Sub_Bass_07_140_C

A second option for creating more rhythmically active bass lines involves the combination of samples with synthesis. When this happens, it creates a hybrid vintage and modern timbre for the project's bass instrumentation. In this scenario, a sample of a bass performance is layered with a sub bass synth, such as the 808 sub bass patch from the second chapter. The vintage bass sound tends to contain more pronounced midrange frequencies, and it cuts through the mix more than the other bass. On the other hand, the modern bass sound tends to reinforce the tone vintage bass, causing it to subtly enlarge the project's low end content.

Audio Example 2.16

In Audio Example 2.16, I demonstrate one method for blending modern and vintage bass styles. From 00:00–00:36, listeners will hear a drum groove with both a modern, heavily distorted 808-style bass, as well as a sample of a vintage bass guitar. From 00:36–01:13, listeners hear this groove again but with the modern bass removed. From 01:13–1:49, I play the groove once more, but this time only the modern bass is present. For the remainder of the example (1:49 on), the original groove plays with both the vintage and modern bass once again (Figure 2.11).

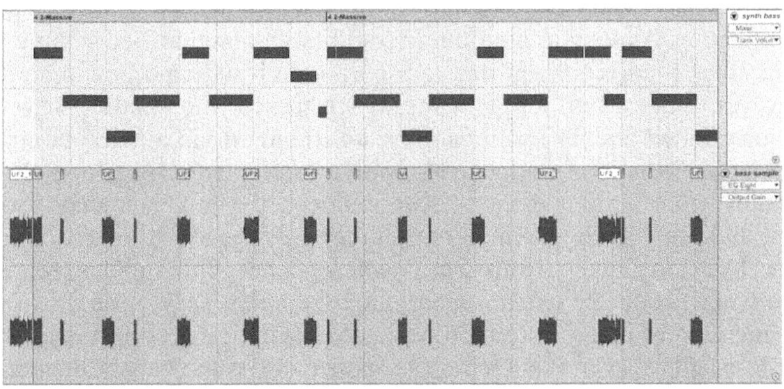

Figure 2.11 A screenshot of the MIDI and audio data of Audio Example 2.16 in Ableton. The bottom (audio) portion is a vintage-sounding bass sample, and the top portion is a modern sounding sub bass synth. Together, these two bass parts form the bassline heard in Audio Example 2.16.

2.4 MELODIC SOUNDS

Melodic samples are the focus of this section. Generally speaking, hip-hop producers tend to use an eclectic array of sounds as melodic material in their tracks. For example, in this genre, it is not uncommon to hear sounds as varied as electric rock guitars, japanese koto, jazz Wurlitzer keyboards, and Indonesian gamelan on the same album. To add interest to these sounds, producers may time stretch, re-tune, slice, or add effects to them in order to suit their aesthetic goals. In some cases, as I discuss in more depth later in this chapter, even non-pitched sounds may serve a melodic function on records in this genre.

As with drum and percussion samples, producers may use continuous (i.e., segments longer than 1 second in length) or non-continuous samples (i.e., segments shorter than one second in length) in order to create melodies for their beats. In the section below, I provide a method for categorizing these sonically wide-reaching sounds using ADSR diagrams, which denote how a sound's attack, decay, sustain, and release (i.e., ADSR) characteristics unfold throughout time. Readers should note that I only categorize short-to-medium length samples this way. And, since these types of files are the easiest to load into a sample-triggering device, CBMP services provide more of them.

Categorizing longer samples in the same way is impossible since any recording longer than 10 seconds or so could technically be referred to as a long duration sample. ADSR diagrams are only useful for shorter duration sounds since they only communicate a sound's amplitude over a short period of time. Thus, these charts are not a useful tool for describing longer audio files.

Instead, I provide an analysis of arrangement strategies for longer samples. Typically, when producers find a longer sample which piques their interest, they tend to reduce this segment into shorter fragments. In many cases, the original (longer) sample will recur semi-regularly throughout the record in order to add emphasis at various points.

Occasionally, these samples are simply used on an "as-is" basis with no modification. As long as a producer combines an unmodified sample with some other sounds, s/he is free to incorporate it within a new recording. However, more often, longer samples are turned into smaller segments, as I mentioned above, and then they are triggered via a MIDI controller such as the Novation Launchpad, Native Instruments Maschine, or one of many other similar devices. Alternatively, the producer may use the point-and-click method, where one simply uses the MIDI matrix editor in their DAW to trigger various samples in sequence. Of course, to reiterate, even when producers use non-continuous samples, they often do so with the intention of creating some type of cohesive digital performance from that instrument, even if such a performance defies the capabilities of any living performer.

After discussing arrangement methods for longer duration samples, I closely examine strategies for arranging short-to-medium duration samples. CBMP platforms provide the most thorough support for these types of samples, even though many longer samples and loops are available. In this section, I cover approaches which pertain to samples of vocals, brass and woodwinds, guitars, strings other and plucked instruments, keys, and FX (i.e., sound effects).

2.4.1 Categorizing Short-to-Medium Length Melodic Samples: ADSR Profiles

A single sample may contain a variety of sounds, as demonstrated in the above example. And, since producers typically modify these samples further by applying editing and mixing techniques, it is most useful to categorize melodic samples based on duration (i.e., continuous versus non-continuous), as described within the previous chapter, as well as the sample's attack, decay, sustain, and release (ADSR) characteristics.

While ADSR filter envelopes are used in synthesizer programming, the concept of ADSR is also useful, in a more general sense, for categorizing diverse sets of sonic phenomena. This is because, in a general sense, ADSR diagrams describe a sound's amplitude characteristics using visual gestures which are easy to comprehend.[1] As pictured in Figure 2.12, a sound's ADSR envelope is a physical description of its amplitude profile over time, once triggered by a MIDI device, sequencer, or DAW. Every sound begins with an onset period called an *attack* period. This segment is followed by a decline in the sound's initial energy, which we describe as a *decay* period. Together, these attack and decay periods are a description

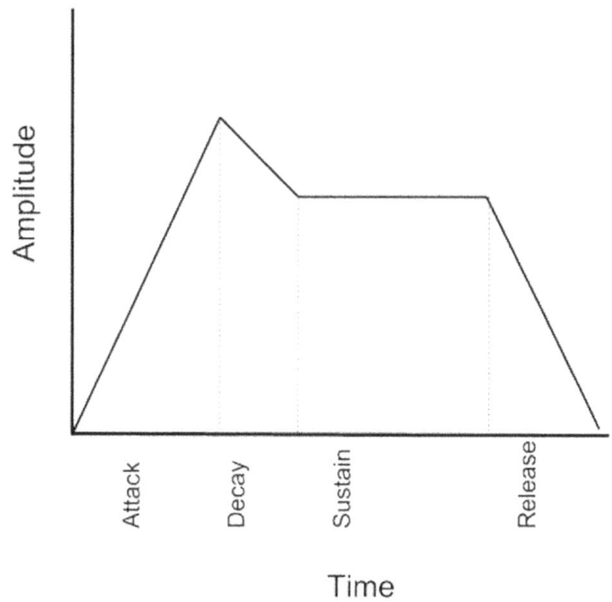

Figure 2.12 An ADSR diagram with a commonly heard envelope.

of a sound's *transient* profile.[2] We refer to the next portion of a sound's ADSR envelope as a *sustain* period, which describes the sound's amplitude activity. Once the sound has finished sustaining, it begins its *release* phase. This term also describes the sound's behaviour once the user disengages a trigger. For instance, if I were to press one of the pads on Native Instruments' Maschine, *release* describes what the sequencer does at the moment I let go of the trigger.

Sounds can be broadly categorized according to their ADSR characteristics. For instance, sounds with a fast attack, fast-to-medium decay time, short sustain period, and quick release produce a *plucked* sound (Figure 2.13). This ADSR profile is representative of instruments such as the guitar, banjo, koto, or pizzicato strings. Sounds with a medium attack (or longer), medium decay (or longer), and long sustain time create a *pad* sound, such as sustained strings or choir vocals (Figure 2.14). Longer attack and release periods sound like a French horn or solo voice which slowly crescendo and then decrescendo into silence (Figure 2.15). Sounds between these two extremes do not have any formal name in the music production world, so I will refer to them as "Medium ADSR" sounds in order to signify that the attack, decay, sustain, or release features of these sounds are somewhere in between the extremes I outline above.

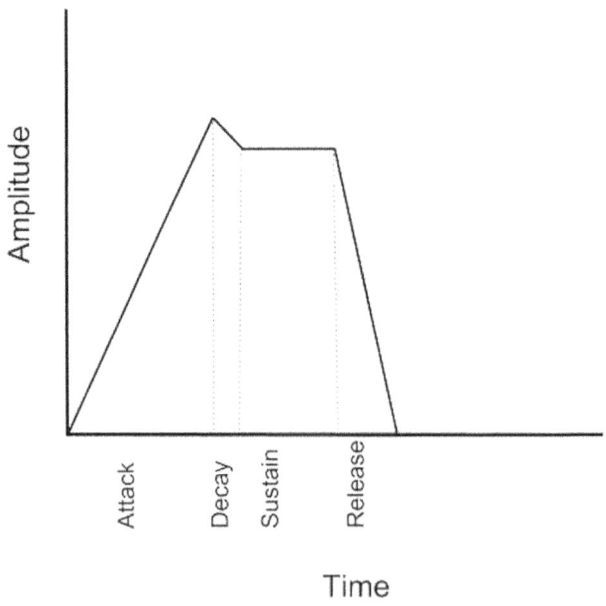

Figure 2.13 The amplitude envelope of a plucked sound.

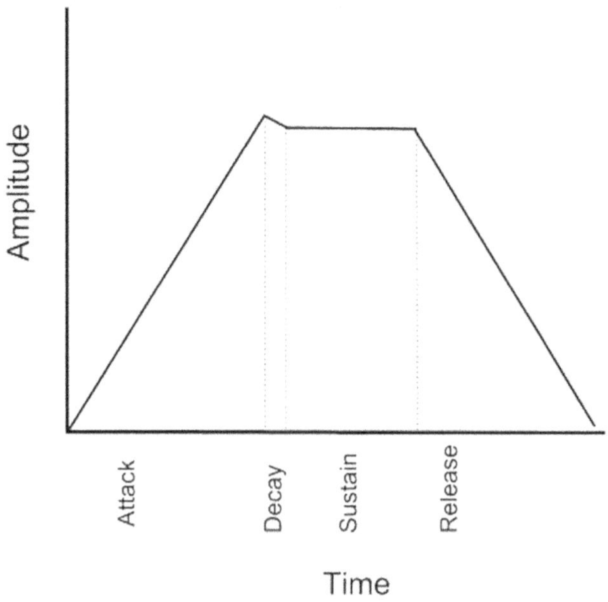

Figure 2.14 The amplitude envelope of a pad sound.

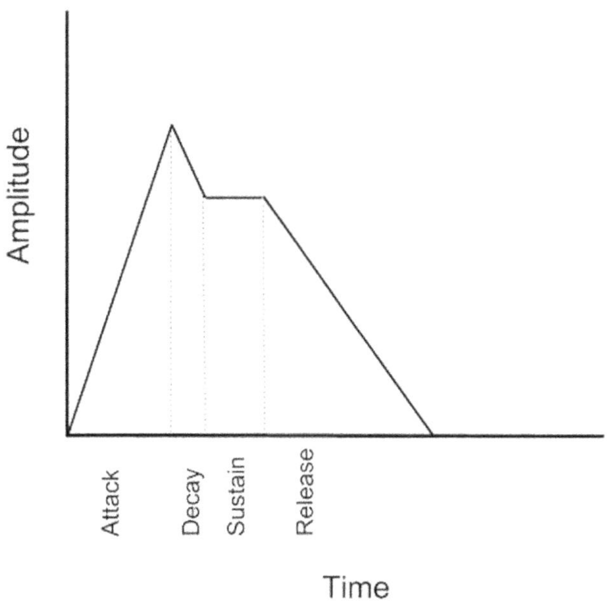

Figure 2.15 The amplitude envelope of an "average" or "in between" sound.

Audio Example 2.17

Audio Example 2.17 demonstrates samples of plucked sounds (Figure 2.13).

Audio Example 2.18

Audio Example 2.18 demonstrates samples of pad sounds (Figure 2.14).

Audio Example 2.19

Audio Example 2.19 demonstrates samples which, in terms of ADSR, have characteristics which are between a pluck and a pad (Figure 2.15).

2.4.2 Arranging Long Duration Melodic Samples

Quite often, when producers turn a continuous sample (such as a break, for example) into smaller slices, they often arrange the track so that the original continuous sample intermittently replays from time to time. When producers use this approach, the longer sample recurs at rhythmically predictable points within the record. For instance, if a sample contains a particularly compelling chord progression, such as a harmonic cadence or a pleasant vocal phrase, the producer might reserve this segment so that it plays on every fourth bar (i.e., bars 4, 8, 12, 16, etc.). In this case, a less compelling portion of the original sample would play during bars 1 through 3. In cases such as this, the overall arrangement and compositional characteristics of the beat are, to varying degrees, structured and constrained by sampling technologies and techniques. As stated earlier in the previous chapter, when producers use sampling to create new musical passages, I use the term *sample phrasing* to describe the creative arrangement and alteration of these samples.

A demonstration of sample phrasing for longer melodic samples can be heard on "Bully Rap" by Sean Price (2012), a track which features a sample of Yigal Bashan's 1969 recording of "Oseh Shalom." On "Bully Rap," from 00:05 until 00:19, a sample of Bashan's original record plays which has been pitch shifted and time stretched. As a result, the sample sounds similar to chipmunk vocals heard on the television cartoon *Alvin and the Chipmunks*. At 00:20, the drums begin to play, and then a small segment of the Bashan sample quietly blends in with the rest of the beat, as a background texture, until 3 full bars (or measures) pass by. This portion of the original Bashan sample only contains subtle strings rather than the chipmunk vocal previously heard within the track. Between 00:29 and 00:31, a segment of the original Bashan sample plays which again features this high-pitched singing. At 00:33, the track returns to the subtle string sample, and this pattern repeats for the duration of the track. Table 2.2 reiterates this information in an easy-to-follow format.

To replicate this technique in their own productions, producers typically begin with a complete or semi-complete groove. From this point, they combine the longer sample with these rhythmic sounds. To do so involves chopping the sample into smaller segments, which last anywhere from a quarter note to 1 full bar or perhaps longer. I demonstrate this approach in the audio example below.

Audio Example 2.20

In Audio Example 2.20, I demonstrate one method for rearranging longer samples in hip-hop beats. In this track, at first, listeners will hear the original unaltered sample. At 00:11, listeners will hear the new version, which I created by slicing the original sample. Each of the slices that I use contain rhythms which stray from the *perfect* digital time grid and, when played in sequence, create a loose rhythmic feel. Listeners will notice that the segment which occurs between 00:19–00:22 (and repeats later on) contains the same ending as the original, unaltered sample.

After the above portion plays, I replay the edited/re-arranged sample with a drumbeat, bass, guitars, FX, and other instruments so that listeners can hear how these edited samples are embedded into a broader musical arrangement.

Table 2.2 Sample Phrasing: "Bully Rap"—Sean Price (2012)

Sonic Event
A pitch-shifted sample from Yigal Bashan's record plays. The sample contains an emotive vocal performance and strings.
Drums begin to play, and a less impactful part of the Bashan sample plays for 3 bars. This segment of "Bully Rap" samples one of the notes played by the strings on the Bashan record.
Three bars after the drums begin, the listener hears an uninterrupted portion of the sampled vocal performance (first heard at 00:05–00:19), which lasts one bar once again.
The sample from 00:20–00:29 plays again.
The sample which begins at 00:29 plays again, with a slight variation.

Unless otherwise noted, in the next section I explain arrangement strategies which pertain to shorter duration melodic sounds. Readers should note that in hip-hop production, as within other forms of electronic music making, the term melodic does not exclusively refer to *pitched* sounds. Instead, for the purpose of this book, melodic sounds refer to prominent sounds which are not immediately identifiable as drums or longer-duration atmospheric and textural sounds (as defined in the Textures, Atmosphere, and Harmony section). As a result, this definition of melodic sounds also includes sound effects (FX).

2.4.3 Arranging Commonly Used Short-to-Medium Duration Melodic Sounds

In hip-hop, many types of melodic sounds can coexist within the same production. In the passage below, I explain (i) various commonly heard melodic sounds, (ii) examples of these sounds in famous productions, and (iii) approaches for using sampling to create these sounds in the context of beat-making. The types of melodic sounds I discuss in this section reflect the categories which CBMP services use to classify samples: vocals; brass and woodwinds; guitars, strings, and other plucked instruments; keys; and FX (i.e., sound effects).

Vocals

It may seem strange to discuss vocals as a component of hip-hop beat creation. After all, hip-hop music is largely based around the presence of an emcee or rapper, whose vocal performances have little to do with the creation of instrumentals. However, vocal samples form the melodic basis of many popular hip-hop beats, such as "Taste" by Tyga and Offset (2018) or "Nice for What" by Drake (2018), which samples Lauryn Hill's "Ex-Factor" (03:25, 1998). Nearly a decade earlier, A$AP Rocky's hits "Wassup" (produced by Clams Casino) and "Peso" from *Long.Live.A$AP* (2011) also make prominent use of vocal samples. Table 2.3 gives a few more examples of hip-hop beats which are based around vocal samples.

As heard on the above records, tracks which are made from vocal samples may consist of series of continuous samples, non-continuous samples, or some combination of both. In terms of ADSR profile, vocal samples typically conform to one of two general curves (Figures 2.16 and 2.17).

For instance, as a result of Statik Selektah's chosen sample arrangement, listeners can hear the lyric "love" being repeatedly sung in "Bird's Eye View" (2013) during 00:05–00:13. In order to create this repeating sound, Statik Selektah (i) sampled a portion of Dorothy Morrison's "Peace Brother, Peace" (1970) before (ii) creating smaller samples from the original larger sample. At 00:00–00:03 and 00:14–00:20, the Morrison sample plays without the interruption or repetition, and, thus, these moments in the song fit the description of a continuous sample provided in this chapter. Between 00:03 and 00:14, listeners can hear shorter samples of Morrison

Table 2.3 Songs Based on Vocal Samples

Song/Artist	*Timing*
"So High" (ft. Currency) by French Montana (2010)	00:00, vocal sample begins
"The Recipe" (ft. Dr. Dre) by Kendrick Lamar (2012)	00:08, vocal sample begins
"Taste" (ft. Offset) by Tyga (2018)	00:00, two vocal samples begin (a low voice and a high voice)
"The Morning" by Raekwon (with Pusha T, Common, 2 Chainz, CyHi the Prynce, Kid CUdi, D'banj) (2012)	00:00, vocal sample begins and then recurs structurally throughout the piece as the hook
"XTCY" by Kanye West (2018)	00:00, vocal sample begins and then recurs structurally throughout the piece as the hook
"Choices (Yup)" by E-40 (2014)	00:02, vocal samples saying "Nope" and "Yup" recur throughout the track
"I Know There's Gonna Be (Good Times)" by Jamie xx, Young THug, Popcaan (2015)	00:02, vocal sample begins song
"Bird's Eye View" by Statik Selektah (2013)	00:00, vocal sample begins, recurs throughout piece
"Wassup" by A$AP Rocky (2011)	00:00, vocal sample begins, recurs throughout piece
"Gucci Gucci" by Kreayshawn (2011)	00:00, vocal sample begins, recurs throughout piece

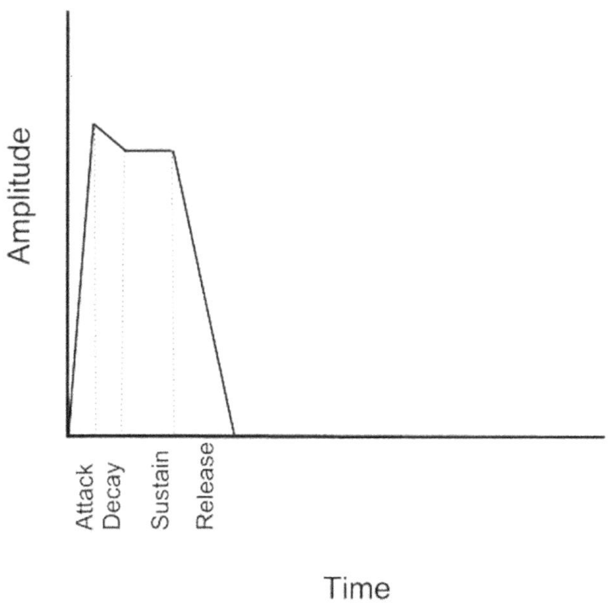

Figure 2.16 The ADSR curve for short vocal sounds and single words like "yeah," "what," and others. Readers will notice that this ADSR curve is the same as the "plucked" ADSR curve from earlier in this chapter.

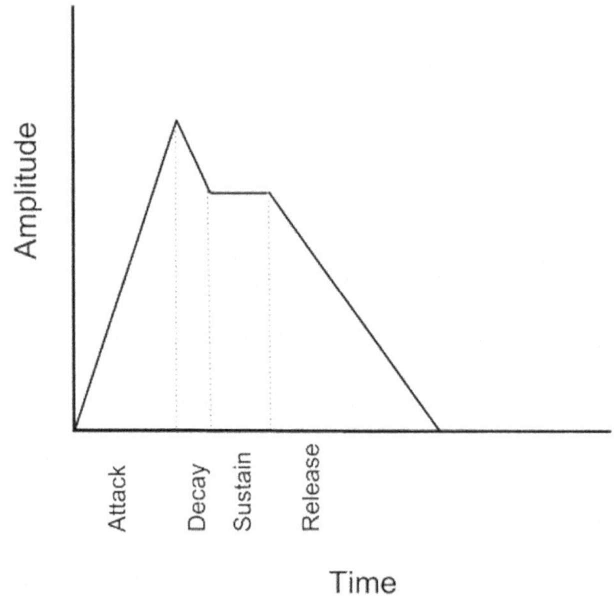

Figure 2.17 The ADSR curve for longer vocal sounds like "oohs," "ahs," and related noises. Readers will notice that this ADSR curve is the same as the "gliss" ADSR curve from earlier in this chapter. If the producer applies enough reverb and delay, the ADSR profile of these sounds can also resemble pads.

singing the lyric "love" over and over again. At 00:14, a new, continuous sample of the Morrison record replays. Readers can follow along with Table 2.4 while listening to this track.

The following audio demonstrations will help listeners identify various ways of arranging vocal melodies in hip-hop beats. For the sake of simplicity, these demonstrations pair various vocal sample arrangement strategies with drums only.

Audio Example 2.21

In Audio Example 2.21, I demonstrate one method for incorporating a continuous vocal sample into a hip-hop beat. At first, a single vocal sample plays. After this occurs, listeners will hear the sample embedded into a hip-hop beat. Please note that the original vocal sample has been downtuned and time stretched.

Vocal sample:
fs_insp90_chill_main_vox2_Am.wav

Audio Example 2.22

In Audio Example 2.22, I demonstrate one method for incorporating a one-shot vocal into a beat. In this case, I used Ableton's Sampler to upload a one-shot vocal sample. This allowed me to map the sample across an entire MIDI keyboard at discrete pitches. Listeners will notice that the sample itself also contains a pitch, which Ableton has replicated and re-pitched across each key.

Vocal sample:
STVC_Vocal_Cut_132_D.wav

Audio Example 2.23

In Audio Example 2.23, I use a method of sample arrangement similar to the style of Statik Selektah's "Bird's Eye View," which I provide an analysis for earlier in this chapter. In this audio example, I take a continuous vocal sample, slice it into smaller segments, and rearrange them. I also introduce the original sample at regular intervals throughout the track.

Listeners will hear an unaltered vocal sample at the beginning of the recording (00:09). Once this sample ends, I introduce a drum track, along with a sliced and rearranged version of the original sample (00:11–00:19). At 00:20, the end of the original sample plays without any additional editing or alteration. The entire section repeats at 00:22.

Quite often, tracks made from vocal samples feature editing which causes the vocal to sound somewhat unnatural yet still recognizable as a human voice. For instance, "Taste" (2018) uses a vocal sample which has been digitally tuned to a lower pitch, whereas the vocal sample in "Nice for What" (2018) contains Lauryn Hill's vocals from "Ex-Factor" (03:25,

Table 2.4 Sample Phrasing/Arrangement: "Bird's Eye View"—Statik Selektah (2013)

Time Stamp	*Sonic Event*
00:00–00:03	Longer Morrison sample plays ("Peace Brother Peace," 1970)
00:03–00:14	Shorter, non-continuous samples of the original Morrison sample play. Each segment lasts approximately 1 second or less. The sample contains Morrison singing the lyric "love."
00:14–00:20	A longer continuous sample of the Morrison record plays.

1998), but this sample has been tuned to a much higher pitch. The vocal sample used on "Wassup" (2011) features heavy delay and reverb. Audio Examples 2.24 and 2.25 demonstrate these techniques.

Audio Example 2.24

In Audio Example 2.24, I demonstrate time-stretched vocals with a simple drumbeat.

Audio Example 2.25

In Audio Example 2.25, I demonstrate pitch-shifted vocals with a simple drumbeat.

Keys: Wurlitzer, Rhodes, and Related Instruments

Electric piano sounds such as Wurlitzers and the Fender Rhodes are commonly heard in hip-hop beats (Figure 2.18). As a genre which finds inspiration in jazz music, while also frequently incorporating samples from this genre, this comes as no surprise. For instance, if one searches for "Chick Corea," the famous electric piano player and bandleader, on Whosampled.com, the search engine returns nearly 100 songs which sample his work.

When producers incorporate samples of these types of electric keyboards, they may either be in the form of one-shot samples or melodic fragments.[3] For instance, one could use a Splice sample such as "srs_wrl80_iguess_FDm.wav" and apply the beat-slicing technique described in this book. This would allow the producer to create smaller one-shots and melodic fragments, from which new music could be made. And, although it may seem strange to do so, it is common to trigger these types of keyboard samples using a keyboard-based MIDI controller (although square pads may also be used).

Since keyboards are percussion instruments, some samples of keyboard performances feature compelling rhythmic content. As a result, producers often work with keyboard samples which are slightly longer than one-shots.[4] For instance, rather than separating a sample at the moments where its transients occur, one could, instead, slice the sample at each quarter note or half note. When working with samples this way, the musical results tend to be quite unpredictable. In particular, the rhythmic outcomes of this technique can sound experimental. This is because when one triggers longer slices of a sample, the resultant sound is constrained by the rhythms present within the original sample (as well as the technologies used to trigger it).

In many situations, this technique will create rhythms which do not perfectly conform to the digital time grid, especially in cases where producers

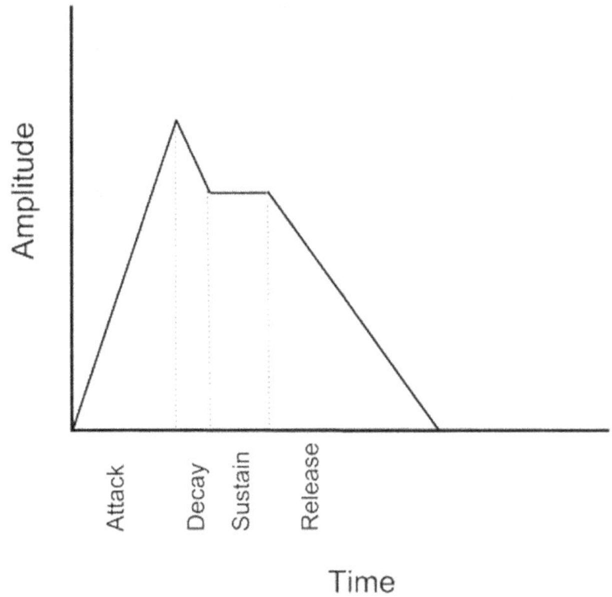

Figure 2.18 The ADSR curve for keyboards like the Rhodes and the Wurlitzer. Readers will notice that this ADSR curve is the same as the "gliss" ADSR curve from earlier in this chapter. If the producer applies enough reverb and delay, the ADSR profile of these sounds can also resemble pads.

use samples which feature bpm fluctuations; a non-consistent metre; and/or performances with *swung* timing attributes, such as the rhythms heard in jazz music. These rhythmic inconsistencies are also known as fluctuations in microrhythm, and this type of off-kilter timing often survives the finalization process for the track's arrangement and mix. In fact, for fans and creators, this is a heavily cherished aspect of the genre (Shelvock 2017a: 175–176).

In the examples below, I demonstrate both the one-shot approach for sampling electric keyboards, as well as a technique which uses melodic fragments. Listeners should note that, as previously demonstrated with vocal samples, incorporating melodic fragments often determines the *phrasing* of the sample within the new production. In this case, the new performance which the producer creates is partially dependent upon on the rhythmic structure of the sample in its original form. When producers add these short recordings to a sampler and record themselves re-triggering these sounds via a MIDI instrument, they also add another layer of rhythmic sophistication to their productions. This is because this technique adds additional timing inconsistencies to a track, which listeners understandably interpret to sound more *human* (Shelvock 2017a: 175–176).

Audio Example 2.26

In Audio Example 2.26, I demonstrate three different approaches for using continuous samples of keyboards in a hip-hop beat. In each instance, first, an unaltered audio sample plays. Once this sample has finished playing, listeners hear a reconfigured version of it, which has been sliced and rearranged using a sample controller. I pair these reconfigured samples with drum loops.

Keyboard samples used (in order of appearance):
CPA_SPK_130_4_Bar_Loop_10_Gm_Vintage
BBL_115_lofi_keys_tremolo_module_Cm
BBL_90_lofi_keys_dusty_fingertips_Dm

Plucked Sounds: Strings, Guitars, and Related Instruments

Another broad category of melodic sounds which find usage within hip-hop beats are strings, guitars, and plucked sounds, which were touched upon earlier in this chapter (Figure 2.19). Generally speaking, plucked instruments include pizzicato strings, guitars, basses, banjos, mandolins, harps, balalaikas, kotos, sitar, ukulele, bouzouki, and many others. Some of these instruments, like strings, may also sustain over a longer period of time. However, I cover this type of sample more thoroughly in the section on Texture, Atmosphere, and Harmony. This is because producers use strings as both plucked instruments (i.e., shorter duration sounds) as well as pads (i.e., long sustaining sounds).

Hip-hop producers incorporate plucked sounds using synthesizer presets which emulate them or, more often, by incorporating samples of the instruments I list in the above paragraph. For instance, "Above the Clouds" by Gangstarr (1998) highlights a sample containing a timbrally ambiguous sounding pluck throughout the entire record. Table 2.5 gives some additional examples.

It is quite easy to slice and trigger plucked samples. This is because plucked sounds have a clear transient profile, and it is easy to identify the onset of their attack and decay periods. Producers can find these samples as one-shots within CBMP libraries, or, alternatively, they may extract shorter audio slices from a longer audio file. Whether a producer slices these samples by selecting custom regions of the sample, or whether they use automatic beat-slicing (as within Logic or Ableton), the clearly defined onset and decay portions of the audio makes them easy to edit and re-trigger.

In terms of arrangement, plucked sounds tend to provide a fairly unobtrusive timbral character. Most often, these sounds occupy middle and high-middle frequencies, allowing them to be clearly heard over denser sonic textures. Since these sounds last for only a short period of time, they also tend to take on a more percussive characteristic.

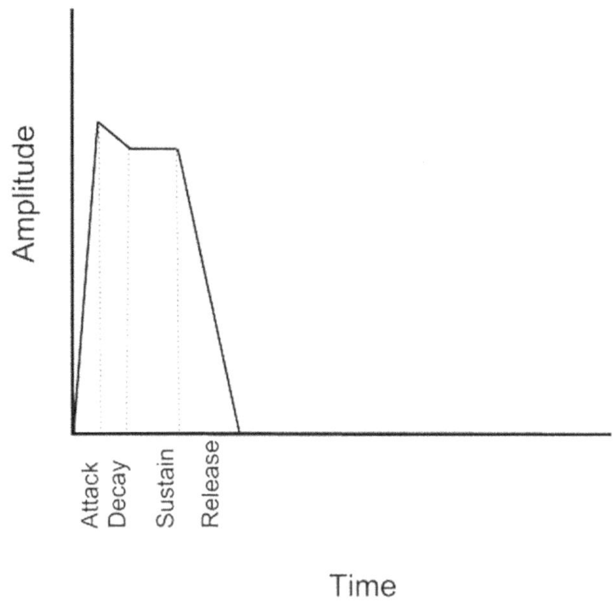

Figure 2.19 Reproduction of the "plucked" ADSR diagram from earlier in this chapter.

Table 2.5 Hip-Hop Releases With Plucked Samples/Sounds

Song/Artist	*Timing*
"My Mind Playin Tricks On Me" by Geto Boys (1991)	00:00, pluck begins and recurs throughout
"Shook Ones, Pt. II" by Mobb Deep (1995)	00:21, pluck begins and recurs throughout
"Street Talkin'" by Slick Rick (1999)	00:00, pluck begins and recurs throughout
"Still D.R.E." by Dr. Dre and Snoop Dogg (2001)	00:01, pluck begins and recurs throughout
"Magic Stick" by Lil Kim (2003)	00:06, pluck begins and recurs throughout
"Sheets" by jhfly (2016)	00:00, pluck begins and recurs throughout
"Devil Eyes" by Hippie Sabotage (2016)	00:00, pluck begins and recurs throughout
"Cadillactica" by Big K.R.I.T. (2014)	00:00, pluck begins (with stereo delay) and recurs intermittently
"Rerocc" by Peewee Longway (2017)	00:00, pluck begins and recurs throughout
"Jazz" by Mick Jenkins (2014)	00:00, pluck begins and recurs throughout

Audio Example 2.27

Audio Example 2.27 is short sample of the song "Rinzai" by my moniker, kingmobb (via ghosttape records, 2017). This track's main melodic sequence was created using a sample called Bpm86_G#_36Chambers_Ethnic_Pluck.wav from the Shaolin Beatz sample pack by Prime Loops. This recording features a performance by an ambiguous stringed instrument which is being plucked by the person playing it. I turned this sample into shorter slices and created a new melody from it.

Brass and Woodwinds

Like with strings, brass and woodwind instruments can find use as either pads (i.e., longer duration sounds) or as plucks (i.e., shorter duration sounds). However, certain instruments in this family, particularly woodwinds, tend to exhibit slightly longer attack, sustain, and release times than typical pluck sounds. Brass instruments, for the most part, tend to offer a punchier delivery, identical to the plucks discussed in the above passage, but some samples may have an ADSR profile more similar to woodwinds with slightly extended attack, sustain, and release phases. I discuss pads more thoroughly in the section titled Textures, Atmosphere, and Harmony.

In trap music, brass horns were exceedingly common in records released between 2012–2016, and this sound still appears in some productions today. These sounds typically last for a short duration, such as heard in some of the examples in Table 2.6.

Brass sounds are also quite common in hip-hop records which feature a Latin influence. Some prominent examples include "Latin Thugs" by Cypress Hill (2004) and "I Like It" by Cardi B (2018). In these cases, the arrangement of the brass instrumentation sounds more like a typical instrumental performance, unlike the trap examples I discuss above. Brass

Table 2.6 Horns in Trap Music

Song Name/Artist	*Timing*
"Original Don" by Major Lazer (Flosstradamus Remix, 2012)	00:40
"Jah No Partial" by Major Lazer (Yellow Claw and Yung Felix Remix, 2013)	01:12
"Love Sosa" by Chief Keef (RL Grime Remix, 2013)	00:56
"Tell Me" by RL Grime and What So Not (2014)	00:54
"Long Distance" by Sam Gellaitry (2015)	00:53

Audio Example 2.28

In Audio Example 2.28, I demonstrate the use of horns in a trap beat. Users can re-create this example using the following examples:

Melodic horn one-shot (to be used with a MIDI controller):
SAC_B_2mo_horn.wav
Bass and drums:
CTS_140_Dm_Future_Trap_Drums_Full.wa
Cinematic horn:
RA_Horn_07_B

sounds may also be highlighted in hip-hop records which do not have a clear Latin influence. For instance, "'93 til Infinity" by Souls of Mischief (1993), "Kick Push" by Lupe Fiasco (2005), "Rigamortis" by Kendrick Lamar (2011), and "King pt 3" by Big Krit (2013) use repeating brass and woodwind samples which use a jazz-based performance approach rather than a Latin one.

Sound Effects (FX)

Sound effects (FX) are as important to hip-hop beats as each of the previously mentioned categories: drums, percussion, and bass; vocals; keys; and guitars, strings, and other plucked sounds. Curiously, in hip-hop production, a sound is classified as an effect based on its function within the arrangement, rather than its timbre. For instance, some sound effects act more like pads, and I discuss these in the section titled Textures, Atmosphere, and Harmony. Shorter duration FX, on the other hand, appear intermittently throughout a project, and I discuss these sounds in this section.

Any sound can be a sound effect, regardless of its timbral characteristics. However, as a generalization, many FX feature oblique sound characteristics in comparison with other sounds on a record. For instance, if a more traditional instrument sound, such as a sax sample, is subject to the application of cavernous reverb and delay, it could be called a sound effect. The sax heard on "93 til Infinity" by Souls of Mischief provides one example of this type of sound effect. This sample plays infrequently throughout the track, typically recurring on beat 1 every 4 bars (00:19).

Readers may notice that I also referred to this example in the section above on brass and woodwinds. I mention this example in both sections because it is, at the same time, a woodwind sample and a sound effect. Brass performances heard on "Kick Push" by Lupe Fiasco (2005), "Rigamortis" by Kendrick Lamar (2011), and "King pt 3" by Big Krit (2013), on the other hand, do not qualify as FX. One reason for this is the instrument performances heard on these records are comprised of continuous samples rather than intermittent ones. In addition, these samples were processed in a more natural-sounding way, unlike the sample heard on "'93 til Infinity." As a result, the samples used in Fiasco, Lamar, and Krit's work are not classified as effects.

CBMP libraries are full of FX samples which producers can search using keywords and filters. This considerably reduces the time it takes find suitable sounds. Users can filter results by (i) arrangement function and/or (ii) the timbre of these sounds. In terms of arrangement function, FX are broadly categorized as (i) transitions (i.e., sweeps and impacts), (ii) atmospheres and textures (as covered in the section Textures, Atmosphere, and Harmony), and (iii) noise. When filtering results by timbre, users may apply a variety of descriptive words. Some filters offered on Splice, for instance, include terms like game, minimal, fantasy, science fiction, reverse, robotic, horror, field recording, and numerous others.

Transition FX help producers connect various sections within a project. If one wants to connect an intro to a verse, or a verse to a chorus, transition effects can help heighten the sense that a new section has begun. The most common transitional effects are (i) sweeps and (ii) impacts. Sweeps are sounds which either rise and fall in amplitude and frequency energy over one or two beats or several bars. CBMP libraries further divide these sounds into (i) risers and (ii) downers, and both of these types tend to last for at least 1 bar. Predictably, risers start out quietly and then increase in amplitude and frequency make-up over time, whereas downers begin loudly and then decrease in loudness and high frequency energy. One of the most common sweep sounds in hip-hop production is the reverse cymbal. Producers can create these samples by reversing ride and crash cymbal samples, or they can simply select a premade reverse cymbal from a CBMP database. Longer duration sweeps, such as risers and downers, are exceedingly common in trap beats and EDM production styles (Table 2.7).

Table 2.7 Sweeps, Risers, and Downers

Song/Artist	*FX Type*	*Timing*
"Red Nation" by The Game, Lil Wayne (2011)	Downer	00:08
"There He Go" by Schoolboy Q (2012)	Downer	00:14
"Choices" by E-40 (2014)	Short riser	00:14
"Hood Gone Love It" (feat. Kendrick Lamar) by Jay Rock (2011)	Short riser	00:14
"Hallelujah" by Mouthe (2016)	Short riser	00:11
"Trees." by Swum (2018)	Sweep	00:06–00:13
"Elephant Man" by So Loki (2018)	Riser	00:29

Audio Example 2.29

Audio Example 2.29 respectively demonstrates 3 sweeps, 3 risers, and 3 downers.

Impacts are transitional effects which tend to occur on the first beat of a new section. For instance, a producer might use an impact when a record transitions from the intro to the verse. To revisit "'93 til Infinity" once again, the iconic saxophone which plays at 00:19 provides one popular example of an impact effect. Quite often, sweeps and impacts are used in combination. For instance, a producer might include a sweep which lasts for 2 bars and then follow up with an impact on beat 1 of the chorus (Table 2.8).

Table 2.8 Impacts

Song/Artist	*Timing*
"S On Your Chest" by Injury Reserve (2016)	00:09
"93 til Infinity" by Souls of Mischief (1993)	00:19
"Devil Eyes" by Hippie Sabotage (2016)	00:33
"Realer Than Most" by Closed Sessions (2014)	00:01
"The Mahdi" by Underachievers (2013)	00:25
"Hood Gone Love It" (feat. Kendrick Lamar) by Jay Rock (2011)	00:15
"Wild for the Night" by Rampage (1997)	1:11

Audio Example 2.30

In Audio Example 2.30, I demonstrate how risers lead into impacts in a typical trap production. This track begins with a vocal sample, snare sample, and a riser. At 00:09, an impact occurs at the moment when the drums and bass begin.

Short duration noise FX are more difficult to categorize than the above sounds. In listening to commercial records, listeners may notice that hip-hop beats contain radio sounds, television and film dialogue, video game noises, found sounds and field recordings, technological noises, environmental sounds, and numerous others (Table 2.9).

Producers typically add various FX sounds to project by loading an assortment of sounds into a triggering device and then recording these sounds using a MIDI instrument interface. FX sounds can add a great deal

Table 2.9 Miscellaneous Sound FX in Hip-Hop

Song/Artist	*Source*	*Timing*
"Peruvian Cocaine" by Immortal Technique (2003)	*Scarface* (1983, Feature Film), monologue	00:00
"Razorblade Salvation" by Jedi Mind Tricks (2006)	*The Holy Mountain* (1973, Feature Film), monologue	03:22
"XTCY" by Kanye West (2018)	*Streetfighter II* (1991, Video Game), sound FX	01:54
"Bring Da Ruckus" by Wu Tang Clan (1993)	*Shaolin and Wu Tang* (1981, Feature Film), monologue	00:00
"Da Mystery of Chessboxin'" by Wu Tang Clan (1993)	*Shaolin and Wu Tang* (1981, Feature Film), monologue	00:01
"Gimme Some More" by Busta Rhymes (1998)	*Psycho* (1960, Feature Film), background music theme	00:00 (throughout)
"It's Mine" (ft. Mobb Deep) by Nas (1999)	*Heat* (1995, Feature Film), monologue	03:11
"Blueprint²" by Jay-Z (2001)	*The Good, the Bad, and the Ugly* (1966, Feature Film), background music	00:00 (throughout)
"Super Lyrical" (ft. Black Thought) by Big Pun (1998)	*Rocky IV* (1985, Feature Film), monologue	01:55
"Death Around the Corner" by 2Pac (1995)	*American Me* (1992, Feature Film), monologue	00:55

of interest to a record, so producers audition various samples while the beat plays. After a take is recorded, one can simply remove or modify any undesirable sounds using their DAW's built in MIDI editor. I demonstrate this process in Audio Example 2.31.

Audio Example 2.31

In Audio Example 2.31, I demonstrate an assortment of miscellaneous effect sounds. At 00:06, I overlay these sounds with a drumbeat.

2.5 TEXTURES, ATMOSPHERE, AND HARMONY

Since sampling is the most commonly used creative technique in hip-hop production, instrumentals in this genre tend to feature a wide pallet of supportive accompaniment sounds. To this effect, hip-hop beats

Figure 2.20 Conceptual diagram, which places atmosphere on the right-hand side and harmony on the left.

tend to feature either a more harmonic (i.e., chordal) approach, a more *atmospheric* approach, or some combination of both of these styles. As a result, it is useful to imagine harmonic and atmospheric sounds in hip-hop beats as though each category exists on opposite ends of a continuum (Figure 2.20).

On one end of the continuum, when hip-hop producers use a harmonic approach in their projects, they use samples which feature chord progressions or some type of harmonic motion (also known as *chord progressions*). And, as explained earlier in the section of this chapter on *sample phrasing*, sometimes producers structure samples so that certain impactful segments of chord progressions reoccur every 4 or 8 bars, such as cadences.[5]

While some tracks use samples based on stronger chordal movements, like the example above, producers may instead choose to feature any type of chord progression they see fit. For instance, a beat might use a sample containing a chord progression which repeats throughout its duration, such as "Regulate" by Warren G (1994). This track samples "I Keep Forgettin' (Every Time You're Near)" by Michael McDonald (1982). Warren G uses a 4-bar loop of this record, wherein the chord progression primarily alternates between i7 and VI7.[6] Table 2.10 lists a number of hip-hop tracks which feature chord progressions which come from samples.

On the opposite end of the continuum, some tracks don't feature any recognizable chord progression, and they use a more atmospheric approach instead. Before the days of easy-access sample libraries, these sounds were often sampled from vinyl records. Today, cloud-based music production services provide easy access to ready-made atmospheric sounds. Or, alternatively, producers can reuse appealing snippets of music or sound within their projects and apply time stretching, pitch correction, editing, and mixing techniques such as equalization, compression, saturation, modulation, and ambient processing in order to establish never-before-heard sonic textures. In hip-hop production, both approaches are equally common.

In electronic music production, atmospheric samples defy any narrow definition. In general terms, they are typically continuous samples which add some type of persistent sound characteristic to a production. These may include, for example, long drone sounds, pads, vinyl crackle, reversed sounds, and numerous other sonic textures which sound like background noise.

Aesthetically speaking, these persistent sounds play a large part in establishing a record's overall mood. However, unlike samples which

Table 2.10 Commercial Hip-Hop Tracks With Chords/Chord Progressions

Song/Artist	*Timing*
"Bonita Applebum" by A Tribe Called Quest (1990)	00:22 (throughout)
"It Was A Good Day" by Ice Cube (1992)	00:00 (throughout)
"Runnin'" by Pharcyde (1995)	00:00 (throughout)
"Feel Me Flow" by Naughty by Nature (1995)	00:00 (throughout)
"Deception" by Blackalicious (1999)	00:00 (throughout)
"Full Clip" by Gang Starr (1999)	00:00 (throughout)
"Street Talkin'" by Slick Rick and Outkast (1999)	00:00 (throughout)
"Right Where You Stand" by Gangstarr and Jadakiss (2003)	00:00 (throughout)
"Hate It or Love It" by The Game (2005)	00:00 (throughout)
"Sunshine" by Atmosphere (2007)	00:00 (throughout)
"Jazz" by Mick Jenkins (2014)	00:00 (throughout)
"Smuckers" by Tyler the Creator (2015)	00:12 (throughout)
"Stark's Reality" by BADBADNOTGOOD (2016)	00:07 (throughout)
"Girls" by Innanet James (2016)	00:00 (throughout)
"Soulful" by L'Indécis (2018)	00:00 (throughout)

Table 2.11 Commercial Hip-Hop Tracks With Atmospheric/Textural Samples

Song/Artist	*Timing*
"I Ain't No Joke" by Eric B. and Rakim (1987)	00:00 (throughout)
"Mama Said Knock You Out" by LL Cool J (1990)	00:00 (throughout)
"The Choice Is Yours" by Black Sheep (1991)	00:00 (throughout)
"Step in the Arena" by Gang Starr (1991)	00:00 (throughout)
"Ms. Fat Booty" by Mos Def (1999)	00:22–01:57
"Me & Nas Bring It Your Hardest" by Slick Rick and Nas (1999)	00:00 (throughout)
"King Piece in the Chess Game" by Slick Rick and Canibus (1999)	00:09 (throughout)
"The Healing Component" by Mick Jenkins (2016)	00:27 (throughout)
"Pillars of Creation" by Blunt One (2018)	00:00 (throughout)
"Trippin" by Smoke Trees (2018)	00:00 (throughout)

feature distinct chord progressions, atmospheric samples are very difficult to describe using standardized musical terminology. This is because atmospheric samples can contain almost any conceivable sound, as long as the sound lasts for a longer duration than a one-shot. As a result, a sound receives categorization as an atmospheric sample depending on its overall function within a project, rather than only its specific timbral properties, dynamic features, or pitch expression. Table 2.11 gives some examples of beats which primarily feature an atmospheric accompaniment style.

While the examples I provide above demonstrate polar ends of the harmony/atmosphere continuum, producers are not restricted to operating this way. In fact, some producers freely incorporate both approaches. Numerous releases from the Chillhop catalog, a well-known instrumental hip-hop label, demonstrate the use of both recognizable chord progressions and persistent sonic atmospheres at once. To accomplish this, producers on their label often add clearly audible vinyl crackle to their productions (i.e., atmosphere), as well as other pads and ambient noises, and they highlight these continuous sounds within the mix. At the same time, they will incorporate samples of commonly heard jazz chord progressions (i.e., harmony) in conjunction with this loud vinyl hiss (Table 2.12).[7]

2.5.1 Ready-Made and Cloud-Based Textures and Atmospheric Samples

Today, cloud-based sample libraries offer samples which already sound as though they have received this type of processing and editing treatment. For example, 91V_TSV_Vocal_Texture_100_Raj__Fm.wav from *Trap Soul Vocals* by 91Vocals on Splice pays homage to old-school sampling techniques.[8] This recording was EQ'd and edited in order to (i) reduce the frequency bandwidth of the recording, (ii) alter the pitch of the vocals, and (iii) slow down the recording. These alterations, performed by engineers at 91Vocals, are intended to sonically resemble classic lo-fi hip-hop sounds.

Table 2.12 Commercial Hip-Hop Tracks With a Combination of Chordal and Atmospheric Support

Song/Artist	*Chordal Timing*	*Atmospheric Timing*
"Hideout" by Tesk (2018)	00:00 (progression begins)	00:00 (vinyl hiss)
"Levitating" by invention_ (2019)	00:32 (progression begins)	00:03 (vinyl crackle and pad)
"Protected" by Swørn (2019)	00:28 (progression begins)	00:00 (vinyl crackle and pad)
"City Lights" by H.1, stream_error (2019)	00:09 (progression begins)	00:21 (distortion and crackle enter with downer effect and drums)
"Patio" by laqueus (2019)	00:00 (progression begins)	00:00 (crackle enters with snare drum played on brushes)
"Laze" by Flovry (2019)	00:10 (first chord begins)	00:00 (vinyl hiss and water noises)
"Faces" by Knowmadic (2019)	00:12 (progression begins)	00:00 (vinyl hiss and pad)
"Don't Know Why" by SwuM (2019)	00:00 (progression begins)	00:00 (vinyl hiss)
"Self Medication" by Blunt One (2019)	00:00 (progression begins)	00:00 (vinyl hiss)
"Blending In" by Nokiaa (2019)	00:11 (progression begins)	00:00 (vinyl hiss, pad, and melodic sample)

They use the following blurb on their site to describe this sample, as well as others in the collection:

> Trap Soul Vocals features hypnotic melodies, wavey textures and zoned out modern rap and RnB vocals. High-end vocal production meets emotive and experimental sound design. Each sample meticulously designed to act as a spark for a new idea or to seamlessly add colour, soul and character. A blissed out collection of hooks, phrases, chops and lo-fi layers. Expect haunting ambient vox, heavily chopped and pitched phrases, clean vocal hooks treated gently with pop polish but then also hard with Trap autotune and reverb chambers.[9]

As stated in the advertising copy in the above passage, 91Vocals offers a library of samples which sound as though they have already been sampled, edited, and processed. And today's commonly used DAWs make it easy for producers to incorporate these types of sounds within their projects. For instance, in Ableton Live, the default settings automatically cause samples to time stretch to the project's BPM settings.

Audio Example 2.32

In Audio Example 2.32, I pair a premade texture sample with a number of additional elements, such as drums, bass, and vocals. The pitch of this sample has been adjusted downward by two semitones. Texture sample:
BBL_90_textured_melodic_clockwork_Am.wav

Audio Example 2.33

In Audio Example 2.33, I pair a premade texture sample with a number of additional elements, such as drums, bass, and vocals. The original sample was recorded at 115 bpm, but I have stretched it to fit at 66 bpm.
Texture sample:
BBL_115_textured_melodic_plateau_Am.wav

2.5.2 Custom Atmospheres and Textures

In addition to incorporating already-made atmospheric sounds within their projects, producers can also create their own. They do this by (i) editing available samples to serve an atmospheric function in the mix, (ii) blending library-based samples together, and (iii) using mix techniques further modify the texture of the atmospheric sound. A key component of this approach involves experimenting with sample selection, as well as sound-modifying techniques such as those associated with editing or mixing.

When producers use editing to create new atmospheres and textures, they do so by applying techniques such as pitch shifting, time stretching, reversal, slicing, or some combination of these methods. However, each one of these techniques tends to behave in an unpredictable way, and, as a result, beat makers tend to spend a long time experimenting with them.

When altering the pitch of an audio sample, for example, its timbral properties also change. The pitch-shifting algorithms made available within popular DAWs cannot disentangle a sound's pitch (also known as fundamental frequency) from its overall sonic colouration. Thus, any attempt to alter a sound's pitch also changes its overall timbre. For instance, if someone increases the pitch of a vocal sample by a perfect 5th (i.e., 7 semitones) or more, the resultant timbre sounds like chipmunks singing or speaking.[10] In Audio Examples 2.34–2.39, I demonstrate some of the unexpected results which occur when time stretching, re-pitching, reversing, and slicing audio to create hip-hop beats.

Audio Example 2.34

In Audio Example 2.34, I pitch shift a segment from a sample of a string performance and incorporate it within a new beat. This segment becomes a texture, like a pad, which supports the keys and drums.

String sample:
JPS_60_A_Glissando_Violin_1.wav

Audio Example 2.35

In Audio Example 2.35, I pitch shift a segment from a sample of a keyboard performance and incorporate it within a new beat. This segment becomes a texture, like a pad, which supports the drums.

Keyboard sample:
100_D_Gliss.wav

Audio Example 2.36

In Audio Example 2.36, I time stretch a segment from a sample of a vocal performance and incorporate it within a new beat. This segment becomes a texture, like a pad, which supports the drums.

Vocal sample:
D_OLIVER_vocal_fx_ahhhhhhh_01_150_A#

Audio Example 2.37

In Audio Example 2.37, I time stretch a segment from a sample of a guitar performance and incorporate it within a new beat. In this case, the portion I use is just a single chord. This segment becomes a texture, like a pad, which supports the drums and other instruments.

Guitar sample:
BBL_115_guitar_winston_Em

Audio Example 2.38

In Audio Example 2.38, I reverse a segment from a sample of a string performance and incorporate it within a new beat. In this case, I also pitch shift the sample down several semitones. This segment becomes a texture, like a pad, which supports the drums and other instruments.

String sample:
TDDC_150_Panther_VoxStrings_A#m

Audio Example 2.39

In Audio Example 2.39, I reverse a segment from a sample of a synthesized choir performance and incorporate it within a new beat. This segment becomes a texture, like a pad, which supports the drums and other instruments.

Synth choir loop:
009_Looped_Choir_Synth_Loop_122bpm_G_-_80SW1_Zenhiser

NOTES

1. Elsewhere, scholars such as Richard Middleton comment upon the usefulness of *gesture* as a unifying idea in music, which can help describe and analyze aesthetics (1993, 179): "This gives a theoretical basis to the idea that 'gesture' occupies a spectrum, with relationships to obvious corporeal movements at one end and neural pulsations at the other. Not only beat and metre, then, but also the micro-physics of intonation, sound-articulation and timbral adjustment: both are parts of the rhythmic ensemble. Moreover, the bottom end of the spectrum can be extended, as the longer periodicities of affective and cognitive movements come into play; the manoeuvres, traverses, outpourings and reflexes associated with corporeal activities, we might say, are transferred to a mental and emotional

setting, where their scale can be expanded, to generate the patterns of phrase-relationships, pitch-contours, harmonic rhythms and so on."

2. When a sound begins, it is typically accompanied by a high energy burst of sound which only lasts for a short duration. This segment of a waveform is known as a *transient*. For example, when a guitar player plucks a string, the amplitude of the pick on the string is a much higher than the proceeding portion of the sound.
3. Sometimes producers may use a longer continuous sample as well. However, this section focuses on shorter duration melodic samples.
4. Producers may also use one-shots, which I also demonstrate in this chapter.
5. In Western music theory, cadences refer to the points in a chord progression which provide the listener with a sense of *resolution* or completeness. The most complete sounding cadence occurs when chord V (i.e., G major, in the key of C) in a given key resolves to chord I (i.e., C major, in the key of C).
6. These are also sometimes written as IIm7 and VImaj7 in order to denote major and minor tetrachords.
7. Sidechain compression can be applied to the vinyl noise in order to increase the sound of the track *pumping*, a term which refers to noticeable increases and decreases in a sound's amplitude over shorter durations of time.
8. This sample is also available at the following web address, accessed August 7, 2018: https://91vocals.com/products/trap-soul-vocals.
9. Reproduced from https://91vocals.com/products/trap-soul-vocals accessed August 7, 2018.
10. In part, this is probably due to association. The main characters from popular cartoons such as *Alvin and the Chipmunks* and *Chip and Dale's Rescue Rangers* always speak and sing with pitch-shifted vocals, as do numerous other cartoon personalities. However, this chipmunk vocal technique is chosen for its timbral results as well as its ability to alter simply alter pitch.

3

The Mix—Conditioning Musical Materials on Records

The term *mixing* finds a number of different applications within music-making communities. For instance, depending on the musical context, mixing can refer to the actions of DJs, live sound engineers, or those who mix or produce records. For the purpose of this book on CBMP, I only use the latter definition. More succinctly, I define mixing as the act of configuring musical materials within a recording project for the purpose of creating the sonic illusion of a musical performance.

Before I proceed, however, readers should note that I use the word mix both as (i) a verb in order to describe the *act* of mixing (i.e., "to mix") and (ii) as a noun to describe the *result* of one's mixing inputs on a record. To clarify this second case further, throughout this section I describe "the mix" on a record, and by this, I refer to a record's final sonic characteristics as establishing by a mixing process of some type. Some producers and engineers also refer to this as a *mix composite*. Throughout this chapter, I refer to those who create mixes as mixers, mix engineers, and producers.

In this portion of the book, I elucidate some of the most crucial aspects of mix-practice as a component of the creation of hip-hop beats. As a result, I do not provide the same level of clarification here as available in publications by Moylan (2014), Hodgson (2010, 2017), Harding (2017), Paterson (2017), Izhaki (2013), Krotz (2017), Shelvock (2014, 2016, 2017a & b), and Nelson (2017). Readers should also be aware that this section does not universally pertain to all types of mixing practices; instead, I focus solely on analyzing how beat producers *hear* the mix, as well as the actions they take to modify a project's mix characteristics based on what they hear.

In order to explain mixing, I begin by clarifying some of the sonic parameters which mix engineers routinely address. I then proceed to elucidate the aesthetic significance of mixing since the sonic characteristics which are determined during mixing often define a record's genre. More specifically, they address the record's (i) spectral distribution, (ii) dynamic contours, (iii) stereo configuration, (iv) ambient design, and (v) noise profile.

3.1 WHAT IS MIXING?

Beyond DJing and live sound, a suitable description of the craft of mixing is provided by audio engineer and researcher Justin Paterson (2017: 77), who refers to mixing music as:

> [T]he adjustment of some aspect of recorded audio to enhance the aesthetic of the end-listener experience. . . . In the broadest sense, mixing has a number of functions: to balance the various musical parts in terms of their volume and spectral content, to correct technical anomalies that have crept into the recording process, and to offer a creative platform upon which to enhance the basic tracks.

Indeed, mixing music involves modifying the amplitude and frequency characteristics of numerous disparate tracks, as well the application of corrective processes in order to *clean up* audible errors within the music. Sometimes these corrective processes are referred to as *editing*. However, in home-studio production, editing and mixing rarely occur as wholly distinct steps, and mix engineers often find themselves editing the tracks within a project. In addition, the creative enhancement Paterson describes above is also equally important to the art of mixing and can involve a number of additional tasks, such as the modification of a sound's ambient and spatial characteristics, saturation levels, dynamic contours, and other related sound-designing activities.

In addition, a distinction is drawn within the world of mix engineering between those who favour software tools and those who favour the combination of hardware and software tools. When engineers mix exclusively with software, this approach is known as working *in-the-box* (Patterson 2017: 77). CBMP services, as a technology which uses digitized computer networks, are typically used in conjunction with an in-the-box mixing approach. However, during mixing, producers may opt to incorporate *out-of-the-box* production hardware, such as tape machines, EQs, compressors, and various types of preamps.

Many modern producers tend to favour an ITB approach. One reason for this is that digital mixing tools provide producers with a higher degree of control over signal data when compared to their analog counterparts. For example, devices such as digital parametric EQs, multiband compressors, and brickwall limiters are generally considered to have changed the sound of all modern records by enabling an unprecedented level of sonic control (Paterson 2017: 79–80). When producers and engineers mix songs comprised of samples (which, of course, were previously recorded and mixed by others), these in-the-box mixing tools can provide the necessary means to *shoehorn* these disparate sounds together. Moreover, in CBMP, in general, in-the-box mixing is a crucial component of the artistic process because during the creation process producers constantly switch between composing, sound-designing, recording, and mixing. For instance, Karriem Riggins (Erykah Badu, J Dilla, Common) shares his own approach to mixing while he makes beats, stating (in Houghton 2017):

> I mix as I go. When the song is complete, we're sending it off for mastering. I know that there's automatic side-chaining with compression and everything, but I like to do a lot of that stuff live; live automation. I work with an SSL Nucleus. It's a new Solid State board that just came out. And yeah, I like to automate. I automate everything, man.

Indeed, as Riggins clarifies in the above passage, mixing occurs in tandem with a number of other creative activities in electronic music production. European hip-hop producer eevee (Chill Hop, Vinyl Digital, Inner-Ocean) also clarifies that she works in a similar way. This is what she has to say about her approach to mixing during beat creation (in Houghton 2017):

> I need to mix and EQ elements as I write/arrange a new track. For example, if I want to make a bassline for a sample, I EQ the bass out of the sample so there is room for a new bass. Also, if I put my kick, snare and hi-hats extension in the beat, I prefer to mix them before because it sounds better. If I have a really hard kick and I didn't mix it before, it sounds really harsh and because of that I can't come into the feel of the beat.

Exactly how an engineer or producer mixes music remains a matter of personal taste. Although, for educational purposes, engineer/researcher Roey Izhaki (2013: 24) suggests that mix engineers might consider some of the following questions as they work:

1. How loud are the instruments in relation to one another?
2. How are the instruments panned?
3. How are the different instruments laid out on the frequency spectrum?
4. How far apart do instruments sound from one another?
5. How much compression was applied on each track?
6. Is there any automation applied?
7. How long are the reverbs?
8. How well defined do the instruments sound?
9. How do different mix aspects change as the song advances?

If any of the above parameters need to be addressed, the mixer will use plugins to suitably modify them. The specific methods and techniques used to make these alterations depend on the engineer's preferences. No matter how an engineer chooses to work, a number of sonic parameters must be addressed by the mix engineer for the record to complete. Broadly speaking, this includes the record's spectral distribution, dynamic contour, stereo configuration, saturation levels, ambient profile, and noise levels (Shelvock 2017b: 18–80).[1] In addition, mix engineers also routinely edit performances, and such edits can alter a track's rhythmic aspects via temporal editing, as well as the expression of pitch through the use of software for corrective tuning. Having covered creative editing more thoroughly in the previous chapter, I will now clarify how beat makers modify a record's sonic parameters during mixing.

3.2 ESTABLISHING GENRE THROUGH MATERIAL MEANS

In general, people who mix records modify their approach according to the genre of the projects they receive. For instance, legendary mixer Eddie Kramer (Led Zeppelin, Bad Company, Jimi Hendrix, the Beatles) shares the following explanation of his approach to mixing records (in White 2006):

> Do I simply push up all the faders and then start to balance from there? No! There are certain building blocks within a mix that to me are of prime importance. What does the style of the song tell you about the direction the mix should take? Is it folk, rock, R&B, pop, rap. . .?

Indeed, when one simply listens to records from different genres, it becomes immediately apparent that each genre demonstrates a different sonic approach. In order to tailor their efforts, mixers spend time analyzing other records in order to discern their sonic qualities. The most important of these qualities include dynamic levels, spectral balance, stereo configuration, ambience levels, and noise levels. Alan Parsons (the Beatles, Pink Floyd, Alan Parsons Project) clarifies his own evaluative process, stating (in White 2006):

I listen to a lot of other people's records at the end of a recording stint to assess fashions in spectral balance—ie. how much bass is fashionable and how bright are people mixing? I hate the level war, incidentally, and always go for dynamics rather than a squashed-up sound to make a track sound loud.

Although Alan Parsons does not work on hip-hop records, his comments demonstrate that his approach involves paying a great deal of attention to the sonic qualities of other records. His comments are not unique in any way, as most audio engineering texts provide a similar perspective because listening is crucial to the process of audio engineering (Katz 2007: 45–52; Owsinski 2008: 7; Cousins & Hepworth-Sawyer 2013: 41–50). However, it can be difficult to reverse engineer the specific methods, techniques, or processes other engineers have used on a given record. For instance, when listening to "Tone's Rap" by BADBADNOTGOOD (2016), one might stop to wonder if the drums were recorded to tape originally, or perhaps the drum signal was sent through a loud-and-distorted analog mixing desk via a process similar to the Black Keys on *Brother* (Simons 2011). Or perhaps they used digital plugins to achieve this effect. Even though this is a less likely scenario, it cannot be completely ruled out. As a result, engineers often personalize their approach to a certain extent according to the ways they imagine other records were mixed. Consciously or unconsciously, those who mix music try to make records sound like other records.

A caveat to this emulative approach, however, is that it is challenging to determine exactly how another record was created. In the absence of some form empirical documentation for a project, engineers and producers can only use guesswork. For this reason, many producers and engineers engage in both active and passive ear training in order to inform their approach to mixing. Passive ear training occurs when someone simply engages in focused listening. The traditional way to do this involves listening to recordings of white noise, where an engineer has filtered or boosted various frequency ranges. The following audio examples exemplify this approach, and readers can also use these examples to perform their own passive ear training.

Audio Example 3.1

In Audio Example 3.1, I demonstrate various frequency ranges using white noise. First, I play white noise with only low frequencies present (20Hz–150Hz; 00:00–00:10), then I repeat this approach with the mid frequencies (200Hz–6kHz; 00:10–00:20) and high frequencies (7kHz–20kHz; 00:20–00:30). Listeners can use these segments to learn to identify low, middle, and high frequencies in a broad sense.

Another method of passive ear-training involves focusing one's attention on the record's various sonic components, such as dynamic levels, spectral balance, stereo configuration, ambience levels, and noise levels. For instance, one might listen to Kendrick Lamar's *Good Kid M.A.A.D City* (2012) and take note of the various vocal effects which occur throughout the album. Budding engineers can also create notes which point to specific moments where a certain sound characteristic is present.[2]

Active ear training occurs when someone focuses their attention on these sonic components while operating production tools of any type. For instance, engineers might apply EQs, compressors, ambient effects, or others to white noise, a previously released record (i.e., a full mix), or individual tracks of some type.

Quite often, even within a single genre, mix engineers may use different approaches. For instance, Young Guru (Jay-Z, Ghostface Killah, Cam'ron, Pete Rock, Method Man) likes to start by treating the instrumental track, as he explains in the following passage (in Tingen 2009):

> I normally take a day for a mix. I first do the beats, taking an hour or two, because in rap and hip-hop the beats are the most important element. I'll then spend some time making the sample sit with that, and then I'll work a couple of hours on the vocals. After that I'll go away and watch TV or play PlayStation or surf the Internet, to give my ears a break. I'll come back in and listen again, and

> mix in the music, and do some more tweaks, and I'll stem it, and will go home.

An alternative perspective is offered by Matthew Weiss (9th Wonder, Sonny Digital, Akon), who explains his preference for highlighting vocals rather than the instrumental tracks (in Weiss 2014):

> I divide the success of a Hip-Hop song into three components: The beat, lyrics, and delivery of those lyrics. Which means two of the three components are about the vocals. I think that [d]ucking the vocals is a big misstep in Hip-Hop. If you go to a live show, it's not uncommon to hear someone do an a cappella rap at some point during the set. This is a make-or-break moment. If the lyrics and delivery are good—the crowd is sold—no beat necessary. If the lyrics or delivery fall short, the crowd is walking out the door. In all honesty, some of the most classic Hip-Hop tunes of all time have so-so production. Vocals are that important. The beat can be "meh" or even non-existent, and people can still like the record.

On a broad scale, these mix engineers use different strategic approaches, yet both have worked on numerous hit records within the same genre. And, somehow, listeners recognize the work of both engineers as hip-hop. In fact, both Weiss and Young Guru have even worked with some of the same rappers, such as 9th Wonder. This is because, in mixing, it is far more important that engineers replicate the sonic characteristics of other hip-hop records, rather than the specific methods and approaches which others use.

Since universally applicable heuristic exists to guide those who mix records, in the sections below I clarify each of the material elements of sound which mix engineers and producers routinely address, as well as the main tools which they use to do so. Readers should follow along with the provided audio examples so that they can begin to recognize these sonic categories and how they are impacted by various tools. These include a record's (i) spectral distribution, (ii) dynamic contours, (iii) stereo configuration, (iv) ambient design, and (v) noise profile.

3.3 SPECTRAL DISTRIBUTION

In audio engineering practices such as mixing, the terms *spectrum* and *frequency* are often used interchangeably. Thus, when engineers refer to either of these, they are discussing the same topic. In addition, these terms have great scientific and aesthetic relevance for audio engineers and researchers—in part because frequency data describes the movement patterns of columns of air. The ear interprets timbre according to the speed at which air moves. As explained by Jay Hodgson (2010: 7):

> Frequency measures the number of times per second that the changes in air pressure which define a soundwave recur, and it

is usually expressed in Hertz (Hz) and kiloHertz (kHz). A soundwave which repeats 20 times per second is said to have a frequency of 20Hz, while a sound wave which recurs 20,000 times per second is said to have a frequency of 20 kHz. Humans are only equipped to hear frequencies within this range. Barring variations in biology, frequencies below 20 Hz and above 20 kHz are actively rejected by the human ear.

Indeed, a sound's frequency distribution determines its audibility. Every complex sound (i.e., any sound which possesses a distinguishable timbre) has both a fundamental frequency, sometimes referred to as f0, and an assortment of *partials*. A sound's fundamental frequency determines its pitch, whereas partials directly impact a sound's timbre, and they cause, for instance, a piano, violin, and voice to all emit distinct sonic characteristics even when they all play the same note. *Harmonics* refer to the whole number multiples of the fundamental frequency (i.e., f0, f1, f2, f3, f4, etc.). Melodic instruments tend to possess an abundance of amplitude at these so-called harmonic frequencies. Quite often, humans interpret sounds with an abundance of *even-numbered* harmonics to sound *warm*. A well-known example of such a sound are tube amplifiers when they are operating at a low-to-medium output level. As one increases the output, however, the distortion which is produced by the tubes clipping (and perhaps the speaker distorting) induces more *odd-numbered* harmonics.

When producers and engineers discuss frequency ranges, they distinguish between low (20Hz–150Hz), middle (200Hz–6kHz), and high frequencies (7kHz–20kHz). Midrange frequencies are further divided into low mids (200–500Hz), mids (500Hz–1kHz), and high mids (2–6kHz; Cousins & Hepworth-Sawyer 2013: 103–105). The following illustrations and audio examples demonstrate each of these using Ableton's native EQ applied to a track containing only white noise. As mentioned above, these types of audio examples can be used to train one's ears to recognize a record's sonic properties.

Audio Example 3.2

In Audio Example 3.2, I demonstrate various frequency ranges using white noise. Once again, I play samples of white noise which highlight low, mid, and high frequencies. However, in this demonstration, I build on Audio Example 3.1 by further dividing the midrange into low mids, mids, and high mids. I present these frequency ranges in the following sequence:

Lows (20Hz–150Hz): 00:00–00:05
Low mids (200–500Hz): 00:05–00:10
Mids (500Hz–1kHz): 00:10–00:15
High mids (2–6kHz): 00:15–00:20
Highs (7kHz–20kHz): 00:20–00:25

Although other sonic dimensions, such as dynamic contour, stereo configuration, saturation levels, ambient design, and noise levels, also receive consideration as timbral qualities, authors and engineers tend to discuss timbre as a function of a sound's general spectral characteristics.[3] This is because when engineers modify almost any sonic parameter within a mix, it causes a measurable change in the record's overall frequency distribution. For instance, if one reduces the gain of a bass synthesizer in relation to the other instruments, the resultant mix will contain less amplitude in the low end of the frequency spectrum. To provide another example, if a mixer places a spring reverb emulator on an auxiliary bus in order to add a sense of ambience to the snare drum, guitars, and synths, it may also cause an increase in amplitude in the project's high frequency content.

Equalizers (EQs) provide the most direct route for modifying a record's timbral qualities. An EQ is a device which can be used to alter a signal's frequency response through the application of filters and amplifiers (Shelvock 2017b: 123). As a result, EQs allow engineers to adjust the amplitude of particular frequencies within a signal (Hodgson 2010: 73).

Initially, EQs were developed in order to address spectral inequalities within communications signals, such as those transmitted through early telegraph and telephone lines. For producers who make extensive use of samples in their productions, EQ serves a similar function today. For instance, imagine the challenges faced by DJ Premier when creating the beat for Mos Def's "Mathematics" (1999), which samples a hit from 1972 called "Baby I'm-a Want You" by the Fatback Band. Premier makes extensive use of the introduction of this older recording, especially the bass and guitar parts. In order to blend comparatively lower fidelity vinyl sounds with more modern sounding drums, hip-hop producers often apply EQ.

Producers use EQs to blend tracks, but they can also use them to make sounds more distinct from one another. Owing to a phenomenon known as frequency masking, humans sometimes have difficulty disentangling sounds which possess similar frequency content (Gelfand 2017: 187). In a project where, for instance, two sounds on two separate tracks mask one another at 500Hz, one might simply use an EQ to lower the gain at 500Hz on one of tracks. If so desired, they might also boost a non-competing frequency range, such as 1kHz, on either track. Audio Example 3.3 demonstrates this tendency.

Audio Example 3.3

In Audio Example 3.3, I demonstrate the sound of two tracks masking one another. At first, the kick and bass drum in this example are difficult to distinguish from one another (00:00–00:05). After a short fade out, I reintroduce the problematic tracks with an EQ dip in the bass between 150–250Hz. Once this occurs, the frequency masking is no longer an issue.

In addition to balancing frequency content, Roey Izhaki notes that the application of EQ also has a creative dimension. Like many recordists, he sees the EQ as a sound-designing tool, stating (in 2013: 205):

> Equalizers used today are not [only] employed to make [sounds within a mix] equal to one another, but to manipulate the frequency content of various mix elements. The frequency virtue of each individual instrument, and how it [sounds within] the overall frequency spectrum of the mix, is a paramount aspect of mixing. Operating an equalizer is an easy affair, but understanding frequencies and how to manipulate them is perhaps the greatest challenge mixing has to offer.

Numerous software EQs find application in today's digital music-making scene. Recordists primarily use 5 different types of EQs: parametric, semi-parametric, graphic, peak, and program (Hodgson 2010: 74). However, the majority of those who produce beats, especially in hip-hop and EDM scenes, simply use the parametric EQs which are readily available in the most common production DAWs, such as Ableton, Logic, and Fruity Loops. Few producers, if any at all, use any of the other available options in modern digital music making. If another solution is required, some CBMP services offer plugins. For instance, Splice users can download free EQ plugins, or they can buy, lease, or freely download a variety of well-known EQs, such as those offered by Waves, Fab Filter, Slate, iZotope, and others.

3.4 DYNAMIC CONTOURS

Dynamics are a crucial feature of all musical performances. In beginner piano lessons, for instance, teachers impart upon their students the importance of playing the instrument with both a loud (*forte*) and soft (*piano*) touch. In physical terms, loud sounds exhibit higher sound pressure levels (SPLs), whereas quiet sounds demonstrate lower SPLs. When a musical phrase is expressed using a higher or lower SPL, it can signify differing musical aesthetics to the listener.

SPLs are also closely associated with a record's overall *loudness* characteristics. To control these levels requires a more sophisticated approach than simply turning gain levels up or down on each track or on the master fader. This is because a sound's perceived loudness levels are more related to its average amplitude than any other sonic feature, including its (i) peak levels or (ii) frequency content. Even though these 2 parameters are still important factors in establishing a sound's overall loudness level, average amplitude remains the most crucial determinant.

On records, raising or lowering these levels requires the use of dynamic range control (DRC) methods, such as compression, limiting, and expansion (Shelvock 2017b: 81–120). Engineers use these techniques to control dynamics on records because all records, in all known media, have a fixed dynamic range. Unlike the environments in which we live, records can be entirely silent, such as Vulfpeck's *Sleepify* (2014), or they can contain sound up to a fixed upper limit of 0 db Full Scale (dbFS). In nature, no hypothetical upper limit exists for dynamic range, whereas the dynamic range of digital recordings is between 96dB to 116dB, depending on whether they use a 16- or 24-bit word length. As a result of these limitations, those who make records must condition the dynamic range of all sounds in a project to conform to these guidelines, and this is accomplished using DRC.

Even though DRC tools serve a practical purpose by modifying a sound's dynamic characteristics to conform to the limitations of the recording media, the application of DRC offers more than just a simple technical support for a record's harmonic, melodic, and rhythmic content. Instead, dynamics are a crucial component of that content, and it is impossible to hear a record without also hearing the dynamic scheme producers and engineers establish. This fact may even frustrate some musical performers, such as vocalists or musicians, who believe that their performances on records have more to do with their own creative inputs, rather than production choices made by engineers and producers.

However, in music production, this simply is not the case. Those who create recordings, whether they call themselves engineers, recordists, producers, or another related term, exert far more agency over the material aspects of a record than musicians, listeners, or anyone else. Of course, dynamic range is one of these key components. During the mixing process, the dynamic range of each and every sound within a mix must be carefully considered. Without exception, recordists are primarily responsible for all sonic characteristics on a record, including dynamics.

Producers and mixers primarily use compressors and limiters in order to condition a record's dynamic contours. These tools act as automatic gain-controlling devices, and they share a number of operational features. As a result, I use the word *compressor* when discussing the general characteristics of these DRC devices in this section. The main technical difference between compressors and limiters is the provision of a fixed ratio (often 10:1 or ∞:1) when limiting a signal. Apart from this, these devices are technologically quite similar, even though they may be used for differing purposes within a mix project. In the section below, I point out instances where it is useful to note the distinction between these two devices.[4]

When applying compression to a signal, producers and mixers typically begin by considering a parameter known as the threshold value. Once a threshold value is set, the compressor monitors the signal via a *sidechain* circuit. When the sidechain detects that the signal's amplitude crosses a given threshold, the compressor automatically attenuates the signal's gain level. An additional parameter, known as ratio, determines the amount of amplitude reduction the compressor applies. A 3:1 ratio produces a signal which, instead of exceeding the designated threshold by 3 dB, will surpass it by only 1 dB. In some cases, where simple attenuation is the goal, recordists may be satisfied with this dynamic alteration and simply move on (Shelvock 2017b: 82).

Audio Example 3.4

In Audio Example 3.4, I demonstrate dynamic range compression applied to a drum sample using various ratio settings. At first, listeners hear the Ableton compressor set to a 2:1 ratio (00:05), followed by 3:1 (00:10), 4:1 (00:15), and 10:1 (00:20).

However, depending on the purpose for applying gain reduction, additional make-up gain can be applied to recover the lost amplitude. This additional make-up gain raises the average amplitude of the signal. Engineers often use compression this way in order to raise a track's RMS level, or what some researchers call a track's body characteristics (Cousins & Hepworth-Sawyer 2013: 69). For example, if 2 dB of attenuation is applied to the signal, the compressor could apply 2 dB of make-up gain as the signal leaves the device's main output stage. This boost occurs after the initial gain reduction has occurred and raises the signal's average amplitude by a full 2 dB (Shelvock 2017b: 83).

Audio Example 3.5

In Audio Example 3.5, I build on Audio Example 3.4 by demonstrating dynamic range compression and make-up gain together. At first, listeners hear a drum track with the Ableton compressor applied to it. I set the compressor to a 2:1 ratio (00:05) with no make-up gain followed by the same ratio with the "lost" make-up gain applied (00:10). I repeat this again with a 4:1 ratio (00:15) and then reapply the lost gain (00:20).

In addition to threshold and ratio, a number of other settings control the behaviour of a compressor as it engages and disengages. *Attack* and *release* parameters establish how quickly and for how long the compressor attenuates a signal once the threshold is crossed. For example, once triggered, a compressor can begin attenuating the signal immediately or after a user-designated delay (often set in milliseconds). Simply put, attack controls the length of time it takes for attenuation to occur once the threshold is surpassed, whereas release determines the length of time the compressor remains engaged (Shelvock 2017b: 83).

Release, on the other hand, controls how quickly the compressor shuts off after the threshold is crossed, and fast release times tend to sound less natural than slower ones (Cousins & Hepworth-Sawyer 2013: 61). Imagine, for instance, a signal that receives 5 dB of attenuation via compression. If the device's gain reduction mechanism disengages too quickly, the resulting audio will feature odd fluctuations in amplitude. This type of quick release compression causes an effect known as pumping. In some cases, such as in hip-hop, pumping may be a desirable sonic effect (Hodgson 2011; Shelvock 2017a: 179). However, this artefact is rarely desirable in other musical circumstances—especially on individual tracks within a mix. When pumping is present on a track, it tends to sound less natural (Izhaki 2013: 269). More frequently, medium release settings are used in mixing. Along these lines, many compressors provide an auto-release feature, and this setting can be quite useful when processing a signal which features widely varying dynamic content (Cousins & Hepworth-Sawyer 2013: 62). Auto-release in DRC refers to an adaptive release setting, wherein release times follow the constantly changing dynamic level of the signal input. Generally, the loudest parts of a signal are more heavily compressed than the quieter components when this feature is engaged (Shelvock 2017b: 84).

Audio Example 3.6 demonstrates slow, fast, and medium attacks (medium release) and slow, fast, and medium releases (medium attack).

Audio Example 3.6

In Audio Example 3.6, I demonstrate various attack and release times applied to a drum loop. At first, I demonstrate fast (00:00), medium (00:05), and slow (00:10) attack times with a medium release setting. Then, I demonstrate fast (00:15), medium (00:20), and slow (00:25) release times with a medium attack setting. The ratio is set to 3:1, and the threshold is set so that the compressor is quite active throughout much of the example so that the attack and release settings are more audible for listeners.

In addition to conditioning the dynamic levels on various tracks within a mix (or the stereo bus), compression can also prevent sounds from audibly overlapping one another. This is easily accomplished using *sidechain compression*, a technique which causes a compressor to attenuate the gain of one track (i.e., signal A) in response to the amplitude level of another track (i.e., signal B). To do so, the compressor's sidechain is set to monitor an external track (signal B), rather than the track which is being compressed (signal A). As a result, when the sidechain circuit detects gain levels which surpass the threshold value on this external track (signal B), it reduces the gain of the track on which it is applied (signal A).

While compressors can make disparate sounds work together within a mix, they are also used creatively in hip-hop as an important aspect of the sound-design process. A common example of this occurs when producers use sidechain compression to intentionally cause sonic artefacts known as *ducking* or *pumping*. This effect causes the creation of clearly audible rhythmic and timbral artefacts within a mix, which Hodgson clarifies, stating (in 2011):

> Hip hop recordists tend to use the side-chain on their compressors to buttress the groove, regularly transforming a pad, or ambience, into a rhythmic upstroke. Flying Lotus (FlyLo), for instance, makes regular and overt use of the side-chain on his compressors for this very purpose. A clear example of the celebrated producer's extroverted approach to the side-chain can be heard on "Tea Leaf Dancers," track one on his debut e.p. for Warp Records. Side-chained in its entirety, every component of "Tea Leaf Dancers", including Andrea Triana's lead vocal, pumps dramatically under the kick drum each time it sounds; and FlyLo does precisely the same on his remix of Madvillain's "Shadows of Tomorrow (feat. Quasimoto)". Lone, a well known British DJ, takes a similar approach on Ecstasy and Friends, side-chaining each track under the kick on all but one cut. Other celebrated champions of the side-chain include J-Dilla and Samiyam, and, to a lesser extent, Madlib, Dibiase, Prefuse 73 and Oh No.

Indeed, the overt use of sidechain compression which Hodgson clarifies in the above passage certainly adds rhythmic sophistication to a project. In Audio Examples 3.7 and 3.8, I demonstrate a number of different ways producers may use this technique.

Audio Example 3.7

In Audio Example 3.7, I demonstrate various approaches for using sidechain compression in hip-hop beats. In the first part of this example, I demonstrate how sidechain compression can disentangle kick and bass parts from one another. At first, listeners hear a drum and bass groove with no sidechain compression applied. After 10 seconds (00:10), I play the same groove again with a compressor applied to the bass. This compressor's sidechain circuit is set to monitor the track containing the kick drum. Each time the kick drum provides a sufficient amount of amplitude, the compressor turns on in order to attenuate the bass.

At 00:20, I play a drum loop and pad together with no sidechain compressor. After 10 seconds (00:30), I play these sounds once again with a compressor applied to the pad. The compressor engages each time the kick drum provides a sufficient amount of amplitude, and this causes the pad to sound as though it is *pumping*.

At 00:40, I play a drum loop with a highly reverberant vocal track. After 10 seconds (00:50), I apply sidechain compression to the auxiliary tracks which contain these FX. I set the compressor so that its sidechain is triggered by the kick drum. This causes the gain of the FX tracks to be attenuated each time the kick drum occurs.

Depending on the compressor's attack and release times, the rhythmic "upstrokes" which a producer creates using this technique often resist strict adherence to a perfect digital time grid. This technique is a key rhythmic feature heard in many hip-hop tracks (Shelvock 2017a: 176; Hodgson 2011). Some sources offer the generalization that hip-hop kick drums tend to occur slightly behind the beat, whereas kick and snare tend to land slightly ahead of the beat. This describes one type of groove made famous by musicians and producers such as J Dilla, Questlove (The Roots), Nas, and A Tribe Called Quest.[5] Producers can reinforce these types of grooves via the application of sidechain compression (Hodgson in 2011). In Audio Example 3.8, I demonstrate how sidechaining can reinforce a track's rhythmic characteristics in this way.

Audio Example 3.8

In Audio Example 3.8, I demonstrate how sidechaining can reinforce a track's rhythmic characteristics. This technique is particularly useful for beats which do not perfectly align to the digital time grid. In this audio example, listeners hear a drum loop with a slightly off-time rhythmic structure blended with a percussion loop which is more *on-the-beat* in comparison.

From 00:00–00:10, I demonstrate the groove with no sidechain compression. From 00:10–00:20, I apply sidechain compression with a quick attack setting and a medium release setting.

The final DRC technique which producers use during the beat-making process is known as *limiting*. Where compressors often find usage within individual tracks within a mix, limiters are more likely to be applied to the stereo bus. Some DAWs, such as Ableton or Fruity Loops, also refer to this as the *master* channel. The difference between a compressor and a limiter is that limiters feature a fixed ratio value, often 10:1 or ∞:1. Brick-wall limiters adopt the latter of these ratios (i.e., ∞:1), and those who produce music using DAWs tend to use these devices to prevent their mixes from surpassing an amplitude level of 0 dbFS. If, for example, a mix had a peak amplitude of +1 dbFS, a limiter could be applied to the master channel in order to reduce this peak to 0 dBFS or lower (Figure 3.1).[6]

Audio Example 3.9

In Audio Example 3.9, I demonstrate how to raise a track's average amplitude using limiting. At first, I play a track with no limiting applied, which peaks near 0 dBFS. Then, at 00:10, I play the same track with a gain boost and heavy limiting applied. Listeners will notice that the track is much louder in the second instance.

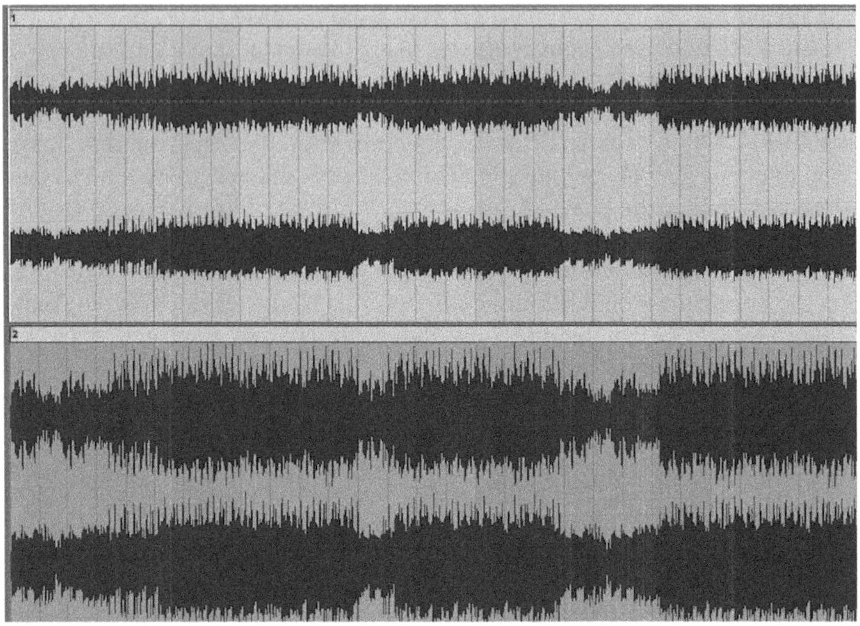

Figure 3.1 The difference between a lower average amplitude waveform (top) versus a higher amplitude waveform (bottom).

3.5 STEREO CONFIGURATION

In mixing literature, stereo configuration generally receives less attention than activities such as EQing and compressing sounds; however, it remains a highly relevant activity. In fact, since the widespread adoption of stereo sound, nearly all records are made with the expectation that they will, at some point, be played using a dedicated 2-channel stereo playback device.[7] Those who mix records modify the stereo field in order to enhance their overall musical impact. As William Moylan explains (2014: 270):

> Music written to be recorded may exploit environmental characteristics, musical balance, and sound quality contradictions, soundstage depth, and width, sound source imaging, or its other unique elements to create, define, or enhance its musical materials.

When discussing records, *width* describes the lateral location of a sound (Moylan 2014: 51). In other words, this term describes the extent to which a sound within a mix sounds as though it is emanating from the left or right extremes of the stereo field.

There are numerous ways to alter the stereo properties of a mix, such as panning, applying the Haas effect, mid/side processing, and applying stereo imaging tools. These techniques can be useful for creating the illusion that certain sounds exist on the *edge* of the stereo field. Or, conversely, tools for stereo width modification can be used to narrow the sound of bass guitar samples and synths, for example, because overly wide bass frequencies (i) are difficult for humans to perceive and, thus, sound much less clear within a mix, and (ii) they can cause an uneven electrical load in playback speakers, which damages speakers at high amplitude levels. In fact, low end frequencies below 300 Hz are typically configured in *mono* format, which, practically speaking, means that there is no difference between the sounds conveyed in the left and right speakers in this frequency range.

For those who make hip-hop beats, stereo configuration techniques are also particularly relevant because these musicians hope to collaborate with vocalists. In order to accommodate an emcee, producers will often leave space in the center of the stereo field. This allows the vocal, which is almost always recorded days, weeks, or years after the beat, to sound more prominent by leaving the center of the mix uncluttered.[8] In hip-hop and other genres, the lead vocal almost always occupies the center of the stereo field (Harding 2017: 62–76).

The simplest technique for altering a sound's stereo placement is known as panning. In every available DAW, users can cause a track to emanate from either the left or right speaker. Most DAWs use a simple graphical depiction of a knob as an interface for moving a sound towards the left or the right portion of the stereo field. Producers and engineers can alter any track within a mix so that it projects from only one speaker or the other, known as *hard left* and *hard right*; however, it is equally common for sounds to be partially moved towards either side.

Audio Example 3.10

Audio Example 3.10 demonstrates speaker panning from left to right.

Engineers and producers can also use a technique known as mid/side (M/S) processing in order redistribute frequency content, or dynamic features, throughout the stereo field. M/S devices allow engineers to control a stereo signal's side and middle portions separately. To accomplish this, they provide discrete access to the *sum* of both the left and right portions of a stereo signal, as well as the *differences* detected between these two components. M/S compression, for instance, can be used to enhance the perceived *power* of the center portion of a mix, and M/S EQ can enhance the perceived "air" (high frequencies) contained in the "sides" of a mix (Cousins & Hepworth-Sawyer 2013: 176). That is to say, M/S EQ allows engineers to control frequencies perceived to emanate from the far left and right regions of a stereo image and to change the stereo balance of a record in the process (Shelvock 2017b: 37).

M/S processing provides engineers and producers with a high degree of control. However, not all engineers prefer this technique. Mix engineer Alastair Sims remarks on his distaste for this technique, when applied to the master stereo bus (in Hepworth-Sawyer & Hodgson, 2017):

> If it's rock then you generally have guitars panned hard left or hard right. If it's Pop then you have a lot of vocals that are panned

> hard left and right or dead up the centre. I spend hours trying to get that balance right. If you just go and crank up the sides then you're turning the guitars up or down and completely changing the balance. That's not what I want.

For Sims, the overapplication of M/S processing on a stereo bus might ruin a carefully balanced array of guitars or vocals.

On the other hand, sample-based music such as hip-hop can often benefit from the application of M/S EQ on individual tracks or the stereo bus. As Grammy-winning hip-hop mix engineer Matthew Weiss states (2016: 5):

> Often times in hip hop and various styles of Electronic music we find ourselves mixing samples—that is, pieces of already completely mixed records.
>
> The difficulty in this is that those samples were mixed as whole arrangements, and now we're ostensibly adding more stuff.
>
> In order to rebalance the samples to make sense with other elements in our new record we sometimes have to get creative with the EQ. And that might mean enhancing a specific guitar sound that lives on the left side of the record while avoiding pulling up the snare, or brightening up the sample in general without turning up the hi-hat that's in there.
>
> M/S can be used to more specifically target what we are trying to accomplish and get more effective results.

Indeed, as Weiss clarifies, incorporating samples into a project can require an extraordinary amount of rebalancing because many samples, whether acquired via CBMP services, CDs, vinyl records, or via another media type, are often full or semi-complete recordings. Thus, they have been mixed, and often mastered, by other engineers or producers and may require a degree of spectral reconfiguration throughout the stereo field. At the very least, engineers often widen sampled material in order to declutter the center of the stereo field so that it does not compete with the emcee's vocal performance (which is normally recorded in a separate session).

Audio Example 3.11

In Audio Example 3.11, I use M/S EQ to disentangle some busy percussion sounds from one another. At 00:00, a busy drum and percussion track plays. At 00:10, I use M/S EQ to (i) mono the lows, (ii) send the high frequencies further outward within the stereo spectrum, and (iii) turn up these high frequencies slightly.

Another more subtle technique than panning and mid/side EQ is the application of the Haas effect. The Haas effect, or precedence effect, occurs when the left or right portion of a stereo signal is delayed between approximately 5–30ms. Rather than creating an echo-style effect, which is a more common usage for delay, the application of this technique causes a track to sound as though it is projecting from either the left or right speaker. In fact, the non-delayed portion of the signal takes psychoacoustic precedence over the delayed portion, and, as a result, the listener perceives the sound that emanate from the non-delayed speaker (Shelvock 2016: 81; Haas 1972: 87).

This technique takes advantage of the human hearing mechanism, which is exceptionally effective at determining a sound's location. When used within a mix, the Haas effect creates an illusory sense of stereo width. However, this technique also creates a sense of depth as well, which can be defined as "the area occupied by the distance of all sound sources" (Moylan 2014: 52).

Although depth is more often associated with a record's ambient characteristics and frequency distribution (as described below in the section titled *Ambient Design*), the Haas effect also establishes a sense of depth. In effect, when a stereo sound is delayed in one speaker, it sounds as though this speaker is slightly further away from the listener. Figure 3.2 illustrates the psychoacoustic impact of this technique. Image A demonstrates how a sound reaches the listeners ears when no such delay is applied. Image B illustrates how listeners hear a stereo signal once a delay has been applied to the right speaker.

Audio Example 3.12

In Audio Example 3.12, I demonstrate the Haas effect. At first, a simple melodic sample plays with no additional effects applied to it. At 00:05, I apply the Haas effect with a setting of 10ms, followed by 15ms (00:10), 20ms (00:15), and 25ms (00:20). Listeners will notice that an increased sense of depth is present in the sound with a delay of 10–20ms, whereas 25ms begins to sound like a discrete echo.

3.6 AMBIENT DESIGN

The ambient properties of a record establish a sense of sonic *depth*, which refers to the perceived proximity of sounds in relation to one another, as well as the listener (Shelvock 2017b: 22). The most commonly used techniques for establishing this quality on records are delay and reverb effects. Although, as discussed above, some depth configuration techniques also cause changes to the stereo field, such as the Haas effect, where time-based processing is also involved.

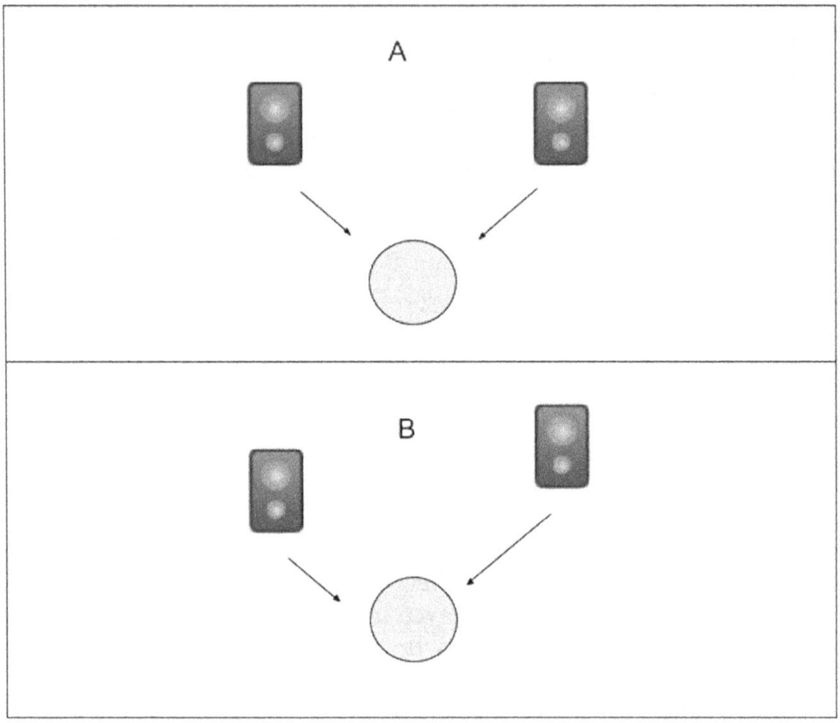

Figure 3.2 The psychoacoustic impact of the Haas effect when it is used in mixing. To create this effect, one simply applies a signal delay to offset the timing in the left or right speaker by up to 30ms. This causes the delayed signal to sound as though it is slightly further away than the non-delayed signal.

Of course, this type depth is purely figurative rather than material (Hodgson 2010: 168). In other words, adding reverb and delay to a track within a mix merely creates an illusion of depth in relation to other tracks in the same mix, as Paul White explains (in 2009):

> In real life we hear sound in three dimensions. Although a stereo production obviously places all the sound sources in front of the listener on a left-right soundstage, it can often seem harder to achieve a good sense of front-to-back depth. . . . Nearby sounds obviously tend to be louder than distant ones, and that's easy enough to mimic, but there are many other factors, too. Consequently, there's a lot more you can do than simply rely on your level faders.

Ambient effects do not increase the measurable depth size of the stereo field in any way. Instead, producers use reverb and delay effects in a way which reminds the listener of other reverberant spaces. Listeners interpret this spatial illusion as sonic depth. In fact, they even have expectations regarding a record's depth characteristics, which are largely informed by the record's genre. Roey Izhaki clarifies this point by stating (in 2013: 72):

> We are used to sonic depth in nature and we want our mixes to recreate such sense.
>
> The main objective is to create something natural or otherwise artificial but appealing.
>
> Our decisions on instrument placement are largely determined by importance and what we are familiar with in nature. We expect the vocals in most rock productions to be the closest, just like a singer is front-most on stage. But in many electronic dance tracks the kick will be in front and the vocals will appear behind it. Sometimes even way behind it. Then we want all drum-kit components to be relatively close on the depth field because this is how they are organized in real life.

Indeed, as clarified by Izhaki, those who mix records rely on the listener's ability to recall what a reverberant or echoey room sounds like. Presumably, people subconsciously store memories of the depth characteristics of various sound environments as they listen to sounds which occur within them (Shelvock 2016: 79). Thus, producers take advantage of the listener's sonic memory, as it were, when creating an illusory sense of depth on a record.

To create this sense of depth, producers use reverb and delay effects. Sometimes, these devices are used in tandem with EQ and some form of spatial reconfiguration (i.e., stereo field alterations). Quite simply, reverb effects emulate the sound of a space or place. Using various hardware and software effects, producers and engineers can emulate the ambient profile of a sound as it reflects and echoes within a given setting. Researcher and engineer Roey Izhaki offers the following perspective on this production technique (in 2013: 405):

> In nature, reverb is observed mostly within enclosed spaces, such as rooms. Reverbs are easier to understand if we imagine an impulse sound, like a hand clap, emitted from a sound source in an empty room. Such a sound will propagate in a spherical fashion and for simplicity we should regard it as traveling in all directions. The emitted sound will travel in a direct path to a listener (or a microphone) followed by reflections that bounced from the walls, floor and ceiling. These will be gradually followed by denser reflections that have bounced many times from many surfaces. As sound both diminishes when traveling through air and being absorbed by surface materials, the reflections will slowly decay in amplitude. Reverb is the collective name given to the sound created by bounced reflections from room boundaries (which we consider to be the main reverb contributors, although in a room there might be many surfaces). In mixing, we use reverb emula-

tors, either hardware units or software plugins, to simulate this natural phenomenon.

Audio Examples 3.13–3.15 demonstrate some commonly heard forms of reverb and delay.

Audio Example 3.13

In Audio Example 3.13, I demonstrate a number of types of delay and reverb in isolation, including a standard digital delay (00:00), a tape delay emulator (00:05), hall reverb emulator (00:10), spring reverb emulator (00:15), and room reverb emulator (00:20).

Quite often, these techniques for altering a record's depth features are used in tandem with techniques for configuring a record's width characteristics. For instance, reverbs often receive use in conjunction with auxiliary sends. A producer can apply reverbs and delays to auxiliary sends and then pan these tracks towards either the left or the right speaker. Another option which producers may use instead of panning involves applying a mid/side stereo imaging device to an auxiliary send. As mentioned above, these tools provide discrete access to the *sum* of both the left and right portions of a stereo signal, as well as the *differences* detected between these two components.

Audio Example 3.14

In Audio Example 3.14, I demonstrate how producers simultaneously address a record's ambient (i.e., depth) characteristics and stereo field configuration (i.e., width). In the first part of the example, a melodic sample plays with no additional delay or reverb. At 00:05, I add reverb to two auxiliary tracks, and I route the track containing the melodic samples to these auxiliary tracks. One auxiliary track is panned 100% the left, and one is panned 100% to the right. The left track has a hall reverb, whereas the right track has a room reverb.

At 00:10, I play the original (unaltered) melodic sample once again so that listeners can reorient their ears to the unprocessed recording. At 00:15, I reintroduce reverb on a single auxiliary send which incorporates M/S processing. I have configured the M/S encoder to localize low and low-mid frequency information to the middle of the stereo field, and I move high mids and high frequencies to the outer portion of the stereo field.

In addition, producers might apply sidechain compression to these auxiliary channels. In this case, the sidechain circuit is set to react to the kick drum in a project. This technique foregrounds the track's groove by (i) making the kick drum sound clearer and by (ii) creating sonic artefacts in response to the kick drum (Hodgson 2011). Each time the kick drum creates sufficient amplitude, the compressor will briefly attenuate the auxiliary channels. This creates the illusion that the kick drum is further forward in the mix because the project's ambient details disappear each time the kick drum plays. In addition, this application creates pumping artefacts in the auxiliary channels each time the compressor restores its lost amplitude.

Audio Example 3.15

In Audio Example 3.15, I demonstrate sidechain compression applied to ambient processing. In this case, I have added drums to the melodic sample presented in 3.14, and I have placed a compressor on the auxiliary sends. The compressor's sidechain circuit monitors the kick drum; when excess amplitude is detected, the compressor briefly attenuates these send channels. Listeners will notice the far left and right portions of the stereo field *pumping* in tandem with the kick drum.

3.7 NOISE, DISTORTION, SATURATION, AND ENHANCEMENT

Distortion and noise seem like they should not belong on a record. However, both of these sonic artefacts find a number of uses within mixing and sound design. In fact, producers use distortion on purpose at times, as Mike Levine of Waves clarifies, stating (in 2018):

> Technically speaking, distortion (aka non-linear distortion) occurs when a waveform's shape is altered between the input and the output of a piece of gear or digital audio plugin or application. This can have both desirable and undesirable effects, depending on how the distortion is created.

Paul White of Sound on Sound elaborates further on commonly heard types of distortion as follows (2010):

> The term "distortion" describes any change made to a waveform: even a simple change to the amplitude is technically a form of distortion, as is EQ, which is, in essence, an amplitude change to specific frequencies. More commonly, though, we intend it to mean

> something that happens when audio passes through a non-linear device, like a saturating tube amp or a clipping preamp. In both cases, the main function is to change the amplitude, but the distortion process introduces new harmonics that are musically related to the original signal in some way.

Indeed, distortion creates additional sound data within a signal which always relates to the input signal. The term *noise*, on the other hand, refers to clicks, pops, and related sonic artefacts that do not seem to belong within a signal. In order to distinguish between these two categories, engineer Bob Katz explains (2007: 139):

> An audio engineer characterizes distortion as a particular form of noise: one that is correlated with the signal. Distortion can be low level and sound much like what is normally called noise, or it can be high level and quite obtrusive, lying on the peaks of the signal.

Noise, like distortion, can be desirable or undesirable within a mix. In fact, on many hip-hop records, producers intentionally include noise from vinyl records such as static, clicks, and pops. If a producer does not actually include any samples of vinyl records in a project, they can add digital samples of vinyl noise instead. Many of these are available on CBMP services, such as the samples in Table 3.1.

Another type of noise which frequently finds its way onto hip-hop instrumentals are environmental recordings, such as street noise, birds, people talking, radio, and other sounds. CBMP services also offer these sounds, such as the examples in Table 3.2.

In addition to noise, many hip-hop records highlight analog and saturation distortion as aesthetic features, and these sounds appear on a track for a number of reasons. Most famously, hip-hop producers sample

Table 3.1 Hiss, Crackle, and Static Examples in CBMP Libraries

Sample Name	*Sample Pack and Creator*
LoFiVinyl.wav	*Aso Cozy Sounds Vol. 1* by Splice
HHI2_97_Vinyl_Lp.wav	*Hip-Hop Instrumentals 2* by Loopmasters
LF_fx_vinyl_noise_02_v2.wav	*Louis Futon's Pretty Cool Sounds* by Splice
CHE_91_Buffering_Crackle.wav	*Hip-Hop Instrumentals* by Loopmasters
Ah_fx_90_howlingcrackle.wav	*Abstract Hip-Hop* by Sample Magic
TDDC_Crackle_06_flexidisc_45rpm_135.wav	*Textures—Dust, Dirt & Crackle Vol. 1* by Capsun ProAudio
BBL_90_crackle_granular_vinyl.wav	B*edroom Beats and Lofi Hip-Hop* by Capsun ProAudio
EK_misc_vinylcrackle.wav	*Ekali Drumkit Vol. 1* by Splice
Decap_texture_vinyl_crackle.wav	*Decap Drums That Knock Vol 5* by Splice

Table 3.2 Environmental and Miscellaneous Noise Examples in CBMP Libraries

Sample Name	*Sample Pack and Creator*
MM_birds_n_shit.wav	Medasin—Microdose Vol 3 by Splice
FW_atmospheric_Ealing_birds.wav	Future Waves and Liquid Electronica by Capsun Pro Audio
BRS_Birds_Seagulls_Lake_Ont_Busy.wav	Essential Birds by Big Room Sound
ForestRain_SO8AM.37.wav	Forest and Woodlands by Blastwave FX
Windstormcold_SO11SSFX.403.wav	Ambience: 5.1 Surround by Blastwave FX
BRS_Forest_Wind_Leaves_Birds.wav	Forest and Woodlands by Blastwave FX
News_Broadcast_Radio_ZA02.407.wav	Horror Vol 2 by Blastwave FX
BRS_Radio_Channel_Change_Bys2.wav	Phones and Mics by Big Room Sound
Playground_ChildrenBWU.13.wav	Parks and Fairs

Table 3.3 Commercial Hip-Hop Tracks With Audible Vinyl Hiss and Crackle

Song/Artist	*Chordal Timing*	*Atmospheric Timing*
"Hideout" by Tesk (2018)	00:00 (progression begins)	00:00 (vinyl hiss)
"Levitating" by invention_ (2019)	00:32 (progression begins)	00:03 (vinyl crackle and pad)
"Protected" by Swørn (2019)	00:28 (progression begins)	00:00 (vinyl crackle and pad)
"City Lights" by H.1, stream_error (2019)	00:09 (progression begins)	00:21 (distortion and crackle enter with downer effect and drums)
"Patio" by laqueus (2019)	00:00 (progression begins)	00:00 (crackle enters with snare drum played on brushes)
"Laze" by Flovry (2019)	00:10 (first chord begins)	00:00 (vinyl hiss and water noises)
"Faces" by Knowmadic (2019)	00:12 (progression begins)	00:00 (vinyl hiss and pad)
"Don't Know Why" by SwuM (2019)	00:00 (progression begins)	00:00 (vinyl hiss)
"Self Medication" by Blunt One (2019)	00:00 (progression begins)	00:00 (vinyl hiss)
"Blending In" by Nokiaa (2019)	00:11 (progression begins)	00:00 (vinyl hiss, pad, and melodic sample)

vinyl records from the 1950s–1970s, which already contain a high level of analog distortion owing to the standard mixing practices of this era (Table 3.3). During this period, it was considered best practice to keep signal levels high, or *hot*, in order to defeat the noise floor associated with analog recording media (Shelvock 2017b: 57). As a result, records released during the vinyl era contain a higher amount of harmonically related distortion (a term that refers to the mathematical relationship between the original signal and the harmonics enhanced within the distorted signal) than typically heard on today's records.

Today, producers may use CBMP samples which feature audible saturation or overdrive, similar to the examples listed above. Alternatively, producers may also apply these effects to *clean*-sounding samples in order to make them resemble classic hip-hop sounds. In the context of digital music creation, when producers want to intentionally create distortion within a signal, they use plugins. These software modules emulate the signal output behaviour of analog circuits. Generally speaking, distortion emulators tend to mimic the sound of vintage overdriven tube amplifiers, channel strips, and other preamps. These plugins are driven by a nostalgia for old sounds, as engineer Hugh Robjohns explains, stating (in 2010):

> In many cases, the technical limitations and imperfections of analogue systems have become an integral part of the quality of the recorded sounds that we all grew up with—and the end result is perceived by many people as being more pleasing than we can easily achieve today with all-digital recording chains. Further than that, some of the sounds resulting from "abuse" of analogue gear have become recognised effects in their own right (tube overdrive and tape saturation being obvious examples).

In the world of CBMP, producers can easily find distortion tools which are offered at no charge. On Splice, for instance, users can freely download distortion plugins such as an 8-bit shaper (Xfer Records), Delta Modulator (Xfer Records), Frohmage (Ohm Force), Tube Amp (Voxengo), and Vinyl (iZotope, Figure 3.3) through the service's various affiliate partnerships.

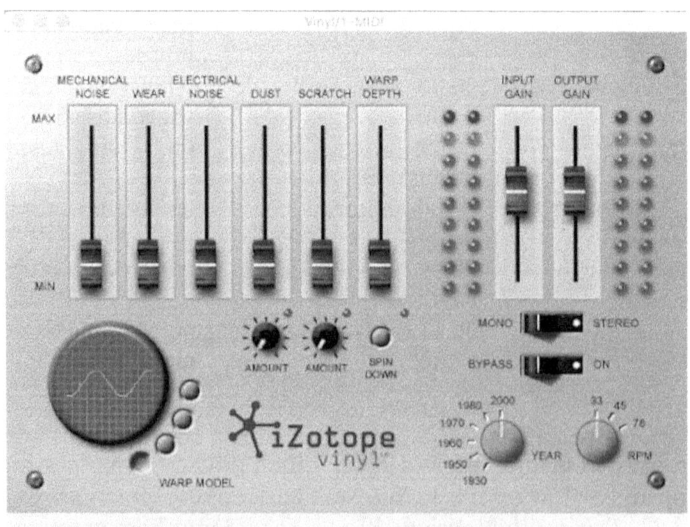

Figure 3.3 Screenshot of iZotope's vinyl plugin. This effect emulates the sound of old vinyl discs playing.

Audio Example 3.16

In Audio Example 3.16, I play a simple (unaltered) drum sample, and I apply various settings from iZotope's vinyl plugin. At first (00:00), the sample plays without this plugin. Then, at 00:05, I apply Setting 1, Setting 2 (00:10), and Setting 3 (00:15) as delineated below.

Setting 1:
Year—1930
RPM—33
Mechanical noise, wear, electrical noise, dust, scratch, warp depth—25%

Setting 2:
Year—1970
Rpm—78
Mechanical noise, wear, electrical noise, dust, scratch, warp depth—25%

Setting 3:
Year—2000
RPM—33
Mechanical noise, wear, electrical noise, dust, scratch, warp depth—25%

Distortion can also be used in a less-noticeable way. One way to do so involves adding overdrive to a sound for the purpose of causing it to appear to move *forward* within a mix.

Stated differently, overdrive can cause sounds to *cut through* a mix by causing the listener to foreground the distorted sounds over the non-distorted sound (Hodgson 2010: 113). By this, I mean that distortion can cause a sound (or sounds) to become more pronounced than other sounds on a record (Shelvock 2017a: 178). Of course, as I discuss above in the section on Spectral Distribution, EQ can also balance sounds within a mix by boosting and filtering select frequencies in order to avoid masking. However, distortion provides a subtler option for accomplishing this.

For instance, if a producer decides to reinforce a synthesizer melody by adding a guitar track which plays the same phrase, one of these sounds could mask some frequency components the other sound.[9] In the very least, since both tracks contain the same pitches, played at the same time, they must possess identical fundamental frequency components. Both sounds will also contain some of the same partials, even though the distribution will differ in both amplitude and frequency to some extent.

Producers can apply distortion to the guitar or synthesizer in order to cause either instrument to sound more prominent within the mix. The most useful types of distortion to apply in this scenario include saturation and overdrive emulators, which can have a subtler effect. These types of plugins generate distortion which is harmonically related to the input signal. This generally creates a more pleasing type of sonic artefact than other distortion techniques, such as bit-crushing, for example (Robjohns 2010).

When mixing hip-hop beats, distortion is commonly applied to either the kick drum or bass instrument(s) in order to cause one of these sounds to sound more prominent than the other. Bass and kick sounds have the tendency to occupy a similar frequency territory, thus causing these sounds to occasionally mask one another. On hip-hop records, the sonic interplay between the kick and bass is a crucial focal point for audiences (Shelvock 2017a: 178), and, as a result, both instruments need to remain audible. Simply turning up the volume fader on one track and turning down the volume of another rarely accomplishes this. Removing the low end from either sound can ruin the mix's overall low frequency balance. However, causing one of these instruments to distort in the low-mid frequencies (i.e., 200–500Hz) or midrange frequencies (i.e., 500Hz–1kHz) can cause the distorted track to sound clearer, or more *forward*, within the mix.

Audio Example 3.17

In Audio Example 3.17, I re-create the masking issue heard in Audio Example 3.3 once again. However, instead of using EQ to separate these instruments, I simply distort the bass in order to make it sound more prominent. At 00:00, listeners hear the problematic samples masking one another. At 00:05, I use the Fabfilter Pro Saturn plugin to distort the midrange frequencies in the bass, causing them to stand out more.

In hip-hop beats, distortion also commonly finds usage on bass instruments. As mentioned in a previous section, the 808 sub bass is crucial to many hip-hop tracks. These instruments nearly always require a degree of distortion in order for listeners to hear them on small speakers, such as laptop speakers and ear buds. Distortion reinforces these bass sounds by generating additional harmonics at higher frequencies, which, in turn, aids the listener in hearing them (Table 3.4).

Table 3.4 Commonly Used Distorted 808 Samples

Sample Name	*Sample Pack and Creator*
SONNY_D_808_2.wav	Sonny Digital Drumkit by Splice
LL_808_F_Crunk.wav	Lex Luger Drumkit by Splice
BQ_808_long.wav	Drum Kit Vol. 1 by Fuse 808 by Splice
KBH_808_crazy_C#.wav	Bighead Sample Pack by Splice
Zay_808_04_E.wav	Zaytoven Drumkit by Splice
FM_808_Legendary_D.wav	Fabian Mazur Sample Pack Vol 1 by Splice
BOI1DA_808_08_C.wav	BOI1DA Soundkit by Splice
JPGC_808_milky.wav	Josh Pan x Gill Chang—Chidlhood Vol 1 by Splice
UZ_808_bt_F.wav	UZ Real Trap Samples Vol 1 by Splice

Audio Example 3.18

In Audio Example 3.18, I take a single 808 bass instrument patch (made using NI's Massive) and distort it a number of ways. Listeners may notice that drums are also present in this example; however, I only cause the 808 bass instrument to distort.

At first, listeners hear the undistorted patch. At 00:05, 00:10, and 00:15, I replay the clip using various settings on the Fabfilter Pro Saturn to distort the 808.

NOTES

1. For the avoidance of confusion, these sonic parameters are addressed once again in audio mastering (please see Shelvock 2017b, *Audio Mastering as a Musical Competency*). However, as it stands, there are few intersections between CBMP and audio mastering, aside from the fact that some services offer one or two mastering tools for free or with a lease-to-own option.
2. Jay Hodgson's *Understanding Records* (2010, 2019) is one of the first texts to offer in-depth listening guides like the ones referred to in this chapter. Readers who are interesting in building skills in mixing should familiarize themselves with the examples in his book.
3. Please see Hodgson (2010), Hepworth-Sawyer and Hodgson (2017), Hepworth-Sawyer, Hodgson, Marrington (forthcoming), and Shelvock (2016, 2017).
4. A treatise on the differences between compressors and limiters far extends the scope of this book. Instead, please see work by authors such as Cousins and Hepworth-Sawyer (2013), Katz (2007), Hodgson (2010, 2011, 2019), and Shelvock (2017), where this topic is discussed in more depth.
5. For more information on this phenomenon, please see Tyler (2018): http://drummagazine.com/drum-lesson-how-to-add-a-questlove-feel-with-the-drag-beats-precise-sloppiness/. Today, producers may follow alternative approaches, such as heard in "Like Dys" (2016) by Anderson Paak and Knxwledge, where the hi hats are slightly behind the beat. In fact, the majority of Suede uses this approach.
6. -.3 dbFS is favoured by mastering engineers. Even though this section discusses mixing, people who produce beats tend to master their beats so that they sound somewhat complete to potential vocalists. This includes ensuring that the beat demonstrates acceptable average and peak amplitude levels.
7. Multichannel music playback (i.e., 5.1 to 22.2 and beyond) has yet to be adopted by the general public in any substantial way, except perhaps the quadraphonic systems of the 1960s and 1970s (which are now obsolete).
8. A common technique for those who mix vocals is to establish a sense of depth and width through the application of stereo manipulation techniques, like those discussed earlier. However, for the most part, the vocal is intended to sound as though it is emanating from the center of the stereo image.
9. While producers often intentionally layer multiple sounds which are timbrally quite different from one another, they might also reinforce a sound by combining multiple instruments with a similar timbre. In this case, each additional layer would have subtle differences.

4

Hearing for Your Audience—Beat Production as a Psychoacoustic Art

In this chapter, I focus on the ontology of beat production as a species of record creation. Record production is a psychoacoustic art form, where one person (or group) encodes an audio document (i.e., a record), and a second person (or group) receives and listens to the same document. Before this document is sent away, the person encoding also listens to it—typically, revising the audio numerous times—before eventually finalizing and distributing it.

In this way, producers hear records *for* listeners, on their behalf, long before these documents are released or sent away. In fact, the record creation process requires producers to constantly scrutinize their work so that they can ensure listeners will correctly interpret the sounds on the record to be musical sounds. The following portion of this book provides an analysis of how the musical materials I discuss so far are used in order to establish a *believable* sounding record for listeners. By believable, I simply mean that producers configure samples, synthesizer recordings, and mixing tools in such a way that the listener accepts them as a document containing music.

To do so, I first explore the role of this type of audible believability which recordists strive to establish on records—*sonic realism* in mixing. Jay Hodgson has referred to records as a "canvas of psychophysiology," wherein specific material and technical conditions must be present in order to for listeners to hear them (Hodgson 2014: 50). Building on Hodgson's ideas, I argue that these psychophysical preconditions directly impact the aesthetics of various production styles, such as hip-hop beat-making.

From this point, I explore Gestalt theory as a model for understanding the perceptual dimensions of the production techniques discussed in this book so far. Numerous designers, artists, scientists, and creators apply a Gestalt methodology in order to analyze the cognitive impacts of various man-made artefacts, such as graphic user interfaces (GUIs), audio interface design, visual illusions, sound designs, and other areas. Gestalt theory even factors into music composition research, such as the writings of James Tenney in *META Meta + Hodos* (1977).

4.1 SONIC REALISM AND THE CANVAS OF PSYCHOPHYSIOLOGY

Those who learn to create records—whether they engage in beat-making or some other type of record production style—inevitably confront some awkward truths about the experience of listening to records. This is because consumers are conditioned to *believe* that the sonic transmissions which occur upon pressing *play* constitute some type of musical performance (Hodgson 2014: 5). When someone begins learning to create these same documents, they must *undo* some of their initial beliefs about how these performances came to be. However, this proves to be difficult since audio engineers and producers spend a great deal of time ensuring that their records do not show evidence of their craftsmanship. If they were to display these efforts in a more obvious way, listeners would have a more difficult time accepting records as *believable* musical performances.

It takes producers a great deal of time and energy to learn to recognize how various production techniques, methods, and approaches create different sound experiences for listeners. In fact, recordists often go to great lengths to create sound documents in a way that seems musically (or at least aesthetically) *realistic*. The specific approach, or *flavour*, of this realism can take many forms, depending on the aesthetic goals of the person who created this auditory document. We typically refer to these differing forms as *genres*.

At first, when musicians learn to create records, it can be difficult for them to disentangle the creative and technical aspects of record production from the illusory musical performance which occurs each time the record is played. As a result, in order to become effective recordists, they train themselves to hear and think like those who create records. This requires the accumulation of skill in (i) hearing and identifying a number of sonic parameters, (ii) understanding how to use the tools which modify these parameters, and (iii) knowing how to alter these sounds so that listeners recognize them as a musical performance.

As I mentioned, there are a number of different ways that a producer might create a believable musical performance on a record. These different flavours of sonic realism are mostly determined by genre, and, to a certain degree, the way a producer likes to work. In some traditions, such as rock music for example, recordists often attempt to re-create realistic-sounding drum and guitar sounds. In rap and hip-hop for instance, we often hear unprocessed and realistic-sounding vocals. However, when these sounds are heard on a record, no matter how *real* they seem, the devices (and people) which made these sounds are not physically present during playback. If the listener happens to believe that the sounds on a record seem realistic, it is only because a recordist created this illusory experience (Hodgson 2014: 5).

At the same time, not all records are created using realistic sounds, which I define to mean sounds that are heard in nature or those which are produced by some acoustic phenomenon. As we often hear in electronic music, hip-hop, and many other genres, many sound FX, synthesized sounds, and edited sounds may strongly resist categorization as *natural* sounds. At the same time, producers and engineers working on these styles of music take great care to ensure that their creations sound *believable* to their listeners in the sense that they sound like music. This type of believability is not to be confused with *accuracy* or *fidelity*. Instead, when listeners believe in a record, they merely accept it as a musical document or artefact. Records which are designed to instill this belief in listeners conform to an aesthetic principle which I call *sonic realism*. I use this term because, although the creation of a record holds almost no resemblance to a live musical performance, we still accept, or *believe in*, the recording as a type of musical performance.

Few records which are intended for commercial release, if any at all, resist this aesthetic tendency (Hodgson 2014: 18). As a result of this deeply embedded expectation for sonic realism on records, for many folks, the transition from simply consuming and enjoying records to creating records requires extensive practice and training. Perhaps the biggest challenge of all, however, is simply learning to *hear* like a producer or sound engineer. Many Focal Press books discuss ear training at length, in addition to numerous other audio-based ear training programs which have been available for many years.

In order to fully hear like a recordist, experts tend to agree that beginners require more than simple auditory exercises. They must also learn to operate production tools at the same time so that they can understand how these studio devices create the sounds heard on records (Katz 2007: 46). Once an audio student addresses both of these aspects at once, they can begin to make inroads towards undoing their conditioned *belief in* records as a representation of some type of acoustic reality (Hodgson 2014: 16). In other words, by gaining skills in these areas, students of audio engineering can remove the curtain concealing the true Wizard of Oz and his tricks, so to speak.

However, most of the time, when researchers and critics analyze records, or when musicians and fans simply discuss them on a casual level, they talk about the sonic characteristics of recordings, in general, as though these sounds are strongly connected to some type of acoustic reality. For instance, scholars, fans, and critics discuss Adele's singing on the record known as "Rollin' in the Deep" (2011) or Eddie Van Halen's guitar technique on "Eruption" (1978) as if these musicians perform for you each time we play these records. But, of course, neither of these musicians are contained within, or trapped inside of, these recordings. More poignantly, these records don't even truly contain actual performances of these musicians. They instead contain data sets, as voltage or binary code, which can be paired with the correct playback device in order to playback a singular auditory perspective on a series of acoustic events which occurred in the past and have since been altered and mutated into a new type of sonic artefact which we call a record (Hodgson 2014: 36).

Although listeners think they are hearing an artist perform on a record, they are actually hearing a single person's perspective on an idyllic music performance which never truly happened. Records provide the listener with a sound document which has already been heard for them, before they ever chose to hit the *play* button. I say this because to make such a document requires the extensive evaluation and modification of sound, including applying various mic techniques, preamp settings (and devices), editing the pitch and timing of musical performances, mixing and arranging the sounds on the record, and perhaps other creative activities as well. At each stage of the record creation process, carefully recorded music performances which occur within meticulously constructed studio spaces are re-evaluated and modified until the producer has decided that the record sounds the way that they would like it to. However, it is a mistake to believe that record itself is *about* any of these performances—records merely present one person's perspective on a series of sonic events (Hodgson 2014: 38).

In fact, primarily due to the marketing efforts of record labels, most listeners tend to confuse the *subject* of a record with *the record itself.* In *Representing Sound*, Jay Hodgson (2014: 6) presents the controversial idea that musicians and performers have far less impact on a record's overall aesthetic than those who create the record:

> Long ago I stopped thinking about performance in the studio as anything other than a sophisticated kind of software programming. When tracking, musicians provide the alternating current—or, in the case of synthesis, the control information—that recordists process, shape, and order into recorded musical communications.
>
> Indeed, performances are the subject of recorded musical communications, whether they are tracked live, synthesized, or sampled. Performances are, in short, the model's pose, the thing recordists portray through their records. In failing to hear records as representations of sound, listeners mistake the subject of communication for communication itself.

The common misunderstanding which Hodgson describes in the above passage is pervasive in music criticism, analysis, and even popular texts. Listeners often confuse representations of sound (i.e., records) with the original sounds being represented on the record. This causes tremendous confusion for those who wish to learn to create records, as well as for those who simply discuss or analyze them in great detail.

For instance, many musicians with little to no knowledge of studio operations tend to believe that recordings are meant to capture and convey the performances of musicians. However, such performances are merely the subject of the record to which they belong. To use analogy from Hodgson, musical performances on records are analogous to "a model's pose" in a photograph. The model's pose is certainly a crucial part of the aesthetic of such a photo, but it only represents one part of an entire artistic project (Hodgson 2014: 8).

In actuality, recordings are highly stylized documents, similar to films. When viewing cinematic recordings, the audience seems much more aware that the actors' performances constitute only one piece of a much larger aesthetic design. Critics and analysts seem to be well aware that an actor's performance is typically the *subject* of the film and not the film itself. However, in music, generally speaking, critics and analysts have not fully accepted this level of sophistication *en masse* in our area of study quite yet.

On recordings of music, the sounds we hear are actually sounds which represent other sounds because sound recording relies on a data storage technique which involves *samples*. Not to be confused with the hip-hop production technique which also goes by the same name, in digital recording to sample is the act of storing snippets of audio data in long sequential chains. A digital hard drive stores samples of a given audio source, such as a vocal performance, which, when played back at the correct speed, replays previously captured sounds. This process resembles the technique of stop-motion animation, as Hodgson (2014: 6) notes in the following passage:

> What we hear when we listen to a record is a sophisticated kind of aural photography, a stop-motion kind of aural animation. Recordists take a series of snapshots, called samples, and synthesists sequence a series of those snapshots, at a rate determined by the recording or sequencing apparatus. Half a century ago, the sample rate required to realistically animate sound was based on the speed at which tape had to pass the playback head of tape machines. That speed was measured in inches per second; typical speeds then were 10 or 15 ips. Likewise, with flat discs and cylinders, the rate was determined by the number of revolutions per minute needed to conjure the same animated illusion, usually 33 1/3 or 45 revolutions per minute. With digital technology like compact discs, digital audiotape, and virtual formats like MP3, WAV, and AIFF, the rate is at least double that at which sounds are sampled—anywhere from 44.1 kHz to 192 kHz.

The discrete samples which Hodgson discusses above are effectively *re-animated* via playback devices, which causes them to sound like a continuous soundstream when they are played back later. Before this happens, the samples are, in essence, heard *for us* already by some type of sound artist or recordist who shapes them. In other words, on a recording, we hear sounds as though they were once ideally heard by a single audio engineer or producer.[1] Indeed, a recording can only offer one sonic perspective on sonic events which have already occurred, and, as a result, producers and engineers often take great care to imagine how others might perceive the same auditory events. Thus, they paint on the canvas of psychophysiology (Hodgson 2014: 47).

However, for the most part, musicians and analysts view the work of those who make recordings, whether they call themselves producers, mixers, recordists, or engineers, as a primarily technical, non-creative, and *non-musical* activity. To compare with film once again, this is like inferring that Orson Welles, director of the classic film *Citizen Kane* (1941), is not an artist *per se* but rather a highly clever technologist. Of course, this is an absurd statement. Yet, many musicians and critics hold on to the idea that the act of creating replayable musical documents is somehow a non-musical endeavour.

4.2 MIXING BEATS: CONSTRUCTING PAST TENSE AURAL NARRATIVES

Many musicians still prefer to think of record production as a technical activity rather than a creative act, even though the accounts of producers, engineers, and even some researchers have shown that this is not the case.[2] Indeed, it's hard to maintain this belief when one simply remembers that a record is a musical document. The sounds contained within any record were put there by a recordist, who took great care to ensure that they conjure some sense of *musicality* for whoever happens to hear the record when it is played.

In order to convince listeners that they are actually hearing *music* when their creations are played back, a number of preconditions must be present. To begin, the communication contained within the record must be recognizable to the listener as a musical communication. Listeners can qualify a number of different sounds and sonic forms as music, from bird songs to classical concertos.[3] No matter what genre, sound, or sonic form is heard on a record, these are all concretized within a *mix*.

The mix is a fairly sophisticated concept. On one hand, when we discuss a track's *mix*, we are talking about all of the sounds heard on a record as a unified soundstream. Sometimes called the mix composite, this soundstream is the sonic ecosystem contained within a record. Every drum, string, voice, crackle, hiss, effect, or noise heard on the record emerges from the mix. On the other hand, to *mix* is also a verb which describes the action of creating a mix. So, the act of altering recorded sounds in almost any capacity can be considered *mixing*. Most often, this includes the signal processing techniques discussed throughout this book. But, in reality, almost any modification one makes to a mix can be considered mixing.[4]

Mixing lays at the heart of all beat-making practices and styles. The art of creating mixes is difficult to learn, just as it is difficult to describe, analyze, and research this creative activity. To make matters worse, the language that musicians and recordists use to discuss mixing, as well as records in general, is over-burdened with unclear terminology. For instance, both live and studio sound engineers engage in the practice of mixing, and, as a result, non-specialists fail to recognize how these practices differ. Even though the cultural artefacts produced by live and studio mixing could not be more different from one another, people with a small amount of skill in either live or studio engineering mistakenly believe that understanding one of these activities automatically provides insight into the other activity. To add to the confusion, many live production and studio-based signal processing tools resemble one another in a broad sense, even though they are used in entirely different ways towards radically different aesthetic goals. Of course, it certainly does not help that, in English, we also clumsily use many of the same terms to describe both live and recorded sounds.

In order to start to address this widespread confusion surrounding records and recording practices, such as mixing, some authors have begun to clarify record production in ontological terms. For instance, one such comes from researcher and professional recordist Jay Hodgson. In *Representing Sounds* (2014: 46), he describes record production as a process rooted in psychoacoustic encoding:

> Records are past-tense aural narratives explaining how a sequence of sounds was once ideally heard, not sequences of sounds per se. They are auditory rather than acoustic narratives—that is, narratives comprising entirely psychoacoustic rather than acoustic information. To understand Recording Practice, then, analysts must first understand the process I call "psychoacoustic encoding." That said, a rather large impediment obstructs our view of that process. Specifically, current conceptualizations of the capacities of sound reproduction technology distort what actually occurs when people make and hear recorded musical communications. . . . [Currently, many assume that] recordists "freeze" acoustic phenomena onto some storage medium . . . and listeners simply reverse the process. But nothing could be further from the truth! Records don't reproduce sounds (i.e., acoustic phenomena). Rather, they model hearing (i.e., auditory phenomena).
>
> As I have already explained, then someone plays a record, they hear something that has already been heard; we know this because the psychoacoustic profiles they hear are fixed, which conjugates sound into the past tense. In fact, the data records contain—a recorded musical communication *in toto*—inscribes particular tympanic motions, simply reversed and fed through electronic circuitry. This data, if used correctly, moves speaker and headphone diaphragms at precise rates in roughly inverse directions and amplitudes that human tympana move to register particular auditory phenomena.

Indeed, records have already been heard for us, even before we hit the play button. Those who create records of any type, whether they call themselves producers, musicians, or engineers, must always mix the sounds they present to us on a record. This requires them to (i) hear the project repeatedly throughout the creation process, (ii) imagine the way they would like it to sound, and then (iii) alter the data encoded within the recording medium to achieve these sounds. Throughout this mixing process, the person creating the record always embeds their own sonic perspective within the mix. From this cognitive vantage point, recordists "paint on the canvas of psychophysiology" the way that they imagine a sound, or collection of sounds, should be best presented for their audiences. Thus, when we listen to a record, we are hearing sounds the way they were once ideally heard by another person. The sounds on a record are heard *for us*, before we even know they exist (Hodgson 2014: 50).

The hypothetical auditor structures the sounds within a mix so that they create a believable auditory narrative. For listeners to accept this narrative, there must be some degree of sonic realism present in the design of a mix—even though the record itself is only an auditory mirage. As I mention in the passage above, sonic realism does not refer to realistic sounds on a record *per se*. Instead, sonic realism refers to the requisite condition than a listener simply *believes in* a record as a sonic transmission. For this to happen, the listener must also suspend disbelief, in an ontological sense, that the sounds s/he hears on a record could never be physically contained within the record itself.

4.3 PAINTING ON THE PSYCHOACOUSTIC CANVAS: APPLYING GESTALT THEORY TO MIXING TECHNIQUE

Building on Hodgson's idea that records are past tense aural narratives painted on a psychophysical canvas, I use Gestalt theory as a tool for analyzing the perceptual impact of mixing techniques (Shelvock 2015, 2016). In other words, Gestalt theory answers the question, "How do we paint on the psychophysical canvas?" In fact, Gestalt theorization has provided artists and critics with tools for analysis within many design-based activities. In particular, Gestalt theory is well suited for the analysis of designs which require perceptual engagement from an end user.

In fact, much of the original Gestalt research which was published in the early 20th century directly investigates the perception of sound in test subjects, even though the theory has been used within a number of other creative and non-creative disciplines. In fact, Gestalt research has a prolific empirical background, and a significant portion of literature on this topic deals with the cognition or perception of sound (Lehar 1999; Wagemans et al. 2012; Moore 2012). In an equally useful sense, available Gestalt research includes a number of studies that demonstrate the general psychophysical tendencies of human hearing.

More germane to the topic at hand, Gestalt theory has also been successfully applied by researchers, scientists, and designers in endeavors such as graphic user interface design, music theory/composition, audio interface design, drawing and painting, and others (Hereford & Winn 1994; Lehar 1999; Wagemans et al. 2012; Moore 2012; Shelvock 2016). This type of literature, which is quite readily available, could serve as a point of departure for future investigations into *why* mixers and sound designers use the techniques they do. However, apart from Hodgson and Shelvock, few resources unpack the psychoacoustic consequences of routine record production tasks (Hodgson 2014; Shelvock 2016, 2017a). In the sections that follow, I use Gestalt theory to describe the psychoacoustic impressions that recordings impart upon listeners when certain mixing techniques are used. In each case, I will describe a common mixing technique, and then

I relate this technique to insights stemming from fields of auditory cognition and perception.

4.4 WHAT IS GESTALT THEORY?

The concept of Gestalt was first introduced by Christian von Ehrenfels in a paper entitled "On Gestalt Qualities" in 1890, wherein he discusses the transpositional properties of musical melodies. In this paper, he notes that a melody can be transposed to new keys using different notes and listeners can still recognize the original melody. He suggests that a *Gestalt quality* structures the perceptual reception of melodies in listeners in such a way that the underlying design characteristics of the melody are preserved when repeated using different pitches. In other words, the melody's constituent pitch relationships create a stronger sonic impression than the specific notes used.

The idea that external stimuli cause perceptual impressions forms the crux of Gestalt theory. Building on von Ehrenfels' work, Wolfgang Kohler introduced the concept of *psychophysical isomorphism*, which explains that there is a correlation between a given external stimuli's design features and the brain's general state while perceiving that stimuli (1938). Generally speaking, this principle works under the notion that properties of consciousness result from accordant electrochemical processes within the brain.

By 1923, Wertheimer elucidated a number of Gestalt theory's most fundamental principles which are often referred to as Gestalt laws (1938). One such principle is the law of Prägnanz, or the law of pithiness, which states that objects within one's perceptual field tend to take on the most simple recognizable structure to the person encountering these objects (Shelvock 2016: 76). Wertheimer also introduces a number of well-known Gestalt laws, such as the laws of proximity, similarity, closure, and good continuation. Most Gestalt research tends to explore these principles.

Researchers in fields such as music theory, composition, and sound design have written at length about the implications of Gestalt theory within their respective fields. For instance, James Tenney's *META Meta (+) Hodos* (1977) posits Gestalt theory as a method for analyzing what he calls *new musical materials* which cannot be easily described via standard notation. In particular, Tenney explains that sonic parameters such as timbre, intensity (i.e., velocity), and rhythmic density are quite important to contemporary music, even though traditional and classical music analysts tend to avoid these topics.

More germane to the topic at hand, sonification researchers Hereford and Winn have also used Gestalt principles in the field known as *human-computer interaction* to design auditory interfaces which are more easily understood by users (1994). Their research explores the creation of *earcons*, which are most easily understood as a type of auditory icon. These auditory icons aid users while they interact with software devices, such as the sounds one hears when using Windows or Mac operating systems

(OS). Winn and Hereford found that Gestalt theory can aid user interface designers by helping them connect sounds to sophisticated computer tasks in a way which users easily understood.

In addition to providing an analytical tool for researchers in music theory, composition, and sound design, Gestalt theorization can find application within the study of music production (Shelvock 2016: 75). On one hand, Gestalt theory can aid researchers in determining the psychophysical impacts of various commonly used production techniques, such as compression, EQ, bus processing, and many others. In addition, Gestalt theory can be used by developing producers and engineers in a predictive way in order to help them consider how to design a mix in a way which satisfies their aesthetic goals.

4.5 AUDITORY STREAM SEGREGATION AND THE LAW OF DISJOINT ALLOCATION

Concerning the analysis of beat-making practices, perhaps the most applicable insight from Gestalt research is the tendency for humans to separate sounds into distinct auditory streams (Moore 2012: 303–313; Miller & Heise 1950: 637–638; Bregman & Campbell 1971: 244–249; Bregman & Rudnicky 1975: 263–267). When listeners are confronted by sounds—whether these sounds are musical or ecological—they perceive the sounds to emanate from either a single source or multiple sources simultaneously (Moore 2012: 303–313).

This tendency towards auditory stream segregation was first demonstrated in a lab experiment by van Noorden in 1975 (1–93). In his dissertation, he explains that a tone labelled "A" was played, followed by tone labelled "B." These tones were played in the sequence ABABAB. The second time around, the sequence was played again with every second "B" eliminated as follows: ABA_ABA_ABA_. Van Noorden's findings demonstrate that, depending on the frequency gap between tones A and B, the test subjects heard either one or two distinct audio streams. When the frequency gap between A and B was smaller, a single stream with a *gallop*-like effect was heard. The instance with the larger frequency gap, on the other hand, causes listeners to separate stream A from stream B as two distinct alternating sound sources.

Broadly speaking, the perceptual illusion of stream segregation is extremely important for mix engineers and beat producers. In fact, it is the job of a mixer to create the sonic illusion that a single stereo auditory stream (i.e., a record) contains several separate streams which sound as though they emerge from various sources. For instance, a record may feature piano, vocals, guitar, and bass. In an ontological sense, each of these sounds emerges from the same sound source (i.e., the record playback device), yet listeners perceive them to form separate soundstreams.

The Gestalt law of disjoint allocation explains this phenomenon. This rule states that natural sounds generally emerge from one source at a time, and, as a result, humans tend to perceive different sounds to originate from different sources. Each source, and the sound which it creates, forms an *audio stream* within the listener's perceptual field. We often perceive complex multi-faceted sounds, like records, which contain only a single audio signal (with left and right components), to contain numerous audio streams at once. On records, this auditory illusion is created via the practices associated with mixing (Moore 2012: 303–313).

This law is demonstrated in an experiment conducted in 1975 by Bregman and Rudnicky (263–267). In their study, tones "A" and "B" are presented alone, and it is easy for listeners to distinguish their order. In the presence of two short distractor tones, however, labelled "x," which occur immediately before "A" and immediately after "B," it is more difficult to distinguish the order of tone "A" and tone "B." The final condition, however, uses a longer series of distractor tones, and once again subjects were able to distinguish tone "A" from tone "B." It was suggested that, given a longer distractor sequence, the distraction tones are perceived to combine into a single perceptual stream and can thus allow tones "A" and "B" to sound distinct in comparison. A shorter distractor sequence, however, causes the "x" streams to separate for listeners, and it is more difficult for tones "A" and "B" to exhibit distinguishable characteristics within this experimental condition (Bregman & Rudnicky 1975: 263–267).

Much of the available Gestalt research literature focuses on the issue of how listeners confront sonic stream separation and stream segregation (Moore 2012: 303–313; Miller & Heise 1950: 637–638; Bregman & Campbell 1971: 244–249; Bregman & Rudnicky 1975: 263–267). I use the term illusion to describe this phenomenon because, in an ontological sense, standard stereo music recordings only exist as a single auditory stream. Although there may be a left and right component, as heard on stereo recordings, the two channels are summed to be perceived as a single sound source (Hodgson 2014: 34). These recordings, as discussed in the above section, entirely rely on this auditory illusion in order to function. Of course, mix engineers take advantage of this sonic mirage in order to cause elements within the mix to sound separate *from*, or cohesive *with*, the other elements.

4.6 LAWS OF SIMILARITY AND COMMON FATE: LOCATION, TIMBRE, AND SIGNAL ROUTING

Gestalt theorists contend that human perceptual mechanisms tend to group similar design elements where possible (Shelvock 2016: 78). For example, Gestalt researchers found that test subjects often grouped similar visual elements such as shapes (Figure 4.1). When applied to audio, the Gestalt rule for similarity suggests that sonic elements are grouped if they possess similar physical dimensions (Moore 2012: 303–313; Miller & Heise 1950: 637–638; Bregman & Campbell 1971: 244–249; Bregman & Rudnicky 1975: 263–267). During mixing, engineers carefully consider how to present the sounds within a mix in terms of similarity and difference. For instance, in order to make a vocal sound more

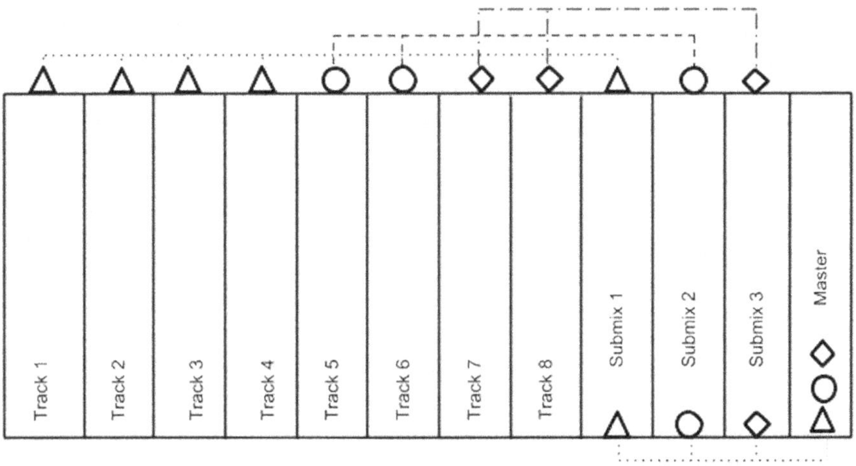

Figure 4.1 Illustrates how subgroups work in mixing. In the above graphic, small shapes at the top of each "track" on the left (triangle, circle, and diamond) correspond with subgroup counterparts on the right (triangle—submix 1; circle—submix 2; diamond—submix 3; all—master channel). Tracks 1–4 would be summed into submix 1. Tracks 5–6 would be sent to submix 2, and tracks 7–8 would be sent to submix 3.

prominent within a mix, some of its physical properties must be different than the other sounds in the mix. However, at the same time, if the vocal sounds *entirely* different than the rest of the mix, listeners may not accept it as a *believable* sound.

Outside of its empirical implications, Gestalt notions of similarity and difference are particularly useful ideas for discussing mixing. To mix a record, recordists consciously process sounds in order to make them sound either cohesive or dissimilar to one another. It's quite easy to imagine why those who mix and produce would want to make certain sounds stand out while perhaps making other sounds less of a focal point. However, these recordists also have many reasons to make sounds somewhat similar to one another. One common method for doing this is called bus processing, for example, which will be discussed in this section.

Three physical properties of sound are particularly important for determining similarities and differences. Humans primarily separate sounds based on differences in fundamental frequencies (F0), differences in timbre over time, and differences in perceived location [10, 17]. In regard to fundamental frequencies, these are, for the most part, chosen before the mixing process. When making beats, the fundamental frequencies of each sound within the mix are typically laid out while choosing drum sounds, programming synths via MIDI, and sequencing parts. On the other hand, modifications to a sound's location within the stereo field, or it's timbral properties throughout the duration of the recording, happen when producers decide to mix their beats.

Panning is the most commonly used mixing technique for designating a sound's location within the stereo field, which allows recordists to cause a sound to emanate from either the left or right speaker or from somewhere in between these extremes. For instance, a track within a mix could be panned 15% to the right or 65% to the left, which essentially means that the sound will favour the left or right speaker to the percentage that the mixer specifies.

Audio Example 4.1

In Audio Example 4.1, I demonstrate the technique known as panning. Panning provides engineers and producers with control over a sound's location within the stereo field. In this example, I pan slowly from the left speaker to the right speaker and then back again.

In addition to panning, those who mix records also use mid/side (M/S) processing devices in order to alter a sound's locational characteristics within a mix. Most often, mixers will use M/S EQ to boost or cut a sound's various frequency components within the center of the stereo field (i.e., the summed version of the signal) or within the sides of the stereo field (i.e., the difference portion of the signal). Recordists may use this technique, for example, to cause acoustic guitar strumming or percussive instruments to sound as though they are projecting from the area near the outside edge of the speakers. This is often accomplished by simply boosting some of the high frequencies by a few dBs in the *sides* portion of the stereo field.

Audio Example 4.2

In Audio Example 4.2, I demonstrate another technique which changes a sound's perceived location within the stereo field known as M/S processing. In this example, a drum loop plays for 5 seconds with no processing added. At this point, the drum loop occurs again with M/S processing to make the lows and low mids more central within the stereo field, whereas the upper mids and mids are more *spread out* between the left and right speakers.

A sound's timbral properties as they unfold throughout time also play an important role in determining its perceptual impact (Moore 2012: 303–313; Miller & Heise 1950: 637–638; Bregman & Campbell 1971: 244–249; Bregman & Rudnicky 1975: 263–267). In other words, human hearing is sensitive to the ways that a sound's timbre transforms from beginning to end. This is why the techniques such as stereo bus processing, subgroup mixing, and parallel auxiliary processing are so important to mixers—they allow mixers to modify the timbre of a number of different sounds at once in the same ways. These techniques take advantage of signal routing possibilities of modern DAWs and mix consoles.

Mixing platforms, whether hardware or software based, allow users to blend sounds together and route them to one common stereo output, or *stereo bus*. However, before the sound reaches this final destination, mixers can control the signal flow in a number of ways through the use of auxiliary buses and subgroups. These different routing options allow users to apply signal processing, such as EQ, compression, reverb, and others, on numerous tracks at once.

The simplest application of this concept occurs when mixers apply a signal processor to the DAW's main output, otherwise known as the stereo bus. By applying EQ to this channel, for example, one can simultaneously change the frequency distribution of all of the tracks within the mix. Although it is considered best practice to achieve the ideal frequency balance by individually adjusting each track, many mixers and beat producers apply subtle stereo bus processing such as EQ, compression, saturation, and other related methods for changing the overall timbral or dynamic qualities of the mix.

Audio Example 4.3

In Audio Example 4.3, I demonstrate the sound of an EQ applied to a stereo bus. This type of approach satisfies the Gestalt Law of Common Fate because all of the sounds within the mix receive an identical treatment. In so doing, a producer or mixer can cause a mix to sound more unified in terms of its spectral content.

For the first 5 seconds of this example, the mix plays with no additional processing. Then, at 00:05, I reintroduce the clip with a +3 dB low end boost (with a wide Q value at 150Hz). At 00:10, I reintroduce the clip with a +3 dB midrange boost (with a wide Q value at 1kHz). At 00:15, I play the clip once more with a boost in the high frequencies (+3 dB using a high shelf filter located at 12kHz).

Audio Example 4.4

In Audio Example 4.4, I demonstrate the sound of a compressor applied to a stereo bus. This type of approach satisfies the Gestalt Law of Common Fate because all of the sounds within the mix receive an identical treatment. In so doing, a producer or mixer can cause a mix to sound more unified in terms of its dynamic content.

For the first 5 seconds of this example, the mix plays with no additional processing. Then, at 00:05, I reintroduce the clip with a compressor which attenuates the peaks in the signal by 1.5 dB. This causes the track's amplitude peaks to sound more uniform from start to finish.

At 00:10, I then add 1.5 dB of make-up gain in order to raise the average amplitude of the track. This causes the track's peaks to be more consistent, while also raising the amplitude of the quieter parts of the track.

Audio Example 4.5

In Audio Example 4.5, I demonstrate the sound of a multiband compressor applied to a stereo bus. This type of approach satisfies the Gestalt Law of Common Fate because all of the sounds within the mix receive an identical treatment. In so doing, a producer or mixer can cause a mix to sound more consistent in terms of the dynamics of various frequency bands.

For the first 5 seconds of this example, the mix plays with no additional processing. Then, at 00:05, I reintroduce the clip with a compressor which attenuates the signal's spectral peaks at 150 Hz (with a wide Q value) by 1.5 dB. This causes the track's low frequency peaks to be reduced consistently from beginning to end, each time the threshold is crossed. At 00:10, I then reverse this maneuver in order to provide 1.5 dB of multiband upward *expansion* at 150 Hz (with a wide Q value). This causes the track's low frequency peaks to increase consistently from beginning to end, each time the threshold is crossed.

Audio Example 4.6

In Audio Example 4.6, I demonstrate the sound of a saturation plugin applied to a stereo bus. This type of approach satisfies the Gestalt Law of Common Fate because all of the sounds within the mix receive an identical treatment. In so doing, a producer or mixer can cause a mix to sound more consistent in terms of the dynamics of various frequency bands.

These Audio Examples cover the application of various signal processors to a DAW's main stereo bus. Another similar method for modifying the timbral qualities of numerous tracks at the same time is to mixing using *subgroups*. To implement this technique, mixers route numerous tracks to subgroups in order to apply some of the techniques demonstrated in the above examples of stereo bus processing. Depending on the number of tracks and the goals of the project, it's common to route all of the tracks within a project to 3–5 submix groups, although more can be used if necessary. These subgroups are routed directly to the DAW's (or mix console's) stereo bus output.

For the most part, subgroups are divided into categories such as drums, instruments, vocals, sound effects, and bass. However, these divisions may change according to the needs of the project, and mixers might also add subgroups for background vocals, strings, and other categories. Figures 4.1 and 4.2 demonstrates the signal flow of subgroups within a DAW or mixing board.

Audio Examples 4.7–4.10 demonstrate the technique of mixing with subgroups in order to modify the sound of the drums in different ways.

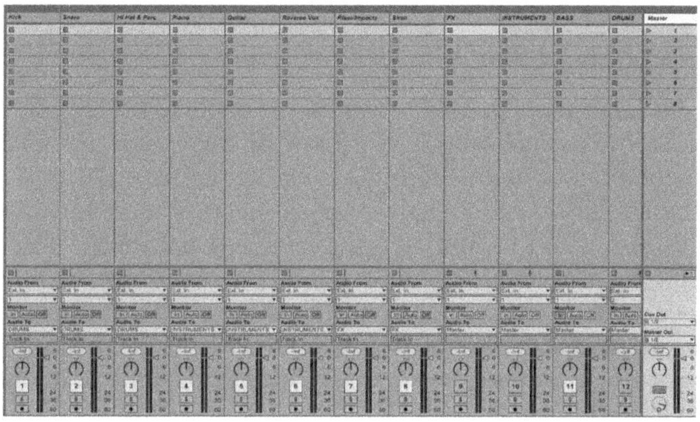

Figure 4.2 Demonstrates one way to set up sub mix groups in Ableton using auxiliary channels. The first 8 channels (left to right) are summed into three channels called "FX," "INSTRUMENTS," and "DRUMS." Readers should note that the channel labelled "Bass" just contains the bass instrument which was used in the project (not pictured).

Audio Example 4.7

In Audio Example 4.7, I demonstrate the application of EQ to a drum subgroup in a mix. In this case, I apply EQ to boost the low and low-mid frequencies of a weak-sounding drum track at 200 Hz by several dB (with a wide Q value). The application of signal processing across numerous tracks at once, as also heard in Audio Examples 4.3–4.6, satisfies the Gestalt Law of Common Fate. This law states that when objects within an observer's (or listener's) perceptual field are altered to resemble one another, these objects tend to be perceived to belong to a larger group.

At first, the example plays with no processing applied to the drum subgroup. After 10 seconds, I reintroduce the clip with the above EQ boost applied.

Audio Example 4.8

In Audio Example 4.8, I demonstrate the application of a compressor to a drum subgroup in a mix. In this case, I apply the compressor to raise the average amplitude of the drum loop. I do so by attenuating the gain of the drum subgroup by 1.5 dB and then by using the compressor's make-up gain feature to add the lost amplitude back into the signal. The application of signal processing across numerous tracks at once, as also heard in Audio Examples 4.3–4.6, satisfies the Gestalt Law of Common Fate. This law states that when objects within an observer's (or listener's) perceptual field are altered to resemble one another, these objects tend to be perceived to belong to a larger group.

At first, the example plays with no processing applied to the drum subgroup. After 10 seconds, I reintroduce the clip with the above compression settings applied.

Audio Example 4.9

In Audio Example 4.9, I demonstrate the application of a multiband compressor to a drum subgroup in a mix. In this case, I apply the multiband compressor to attenuate the mids and upper mids by around 2 dB. The application of signal processing across numerous tracks at once, as also heard in Audio Examples 4.3–4.6, satisfies the Gestalt Law of Common Fate. This law states that when objects within an observer's (or listener's) perceptual field are altered to resemble one another, these objects tend to be perceived to belong to a larger group.

At first, the example plays with no processing applied to the drum subgroup. After 10 seconds, I reintroduce the clip with the above multiband compression settings applied. At this point, I also raise the overall amplitude of the track by 2 dB, thus causing the low frequencies to stand out more than before.

Audio Example 4.10

In Audio Example 4.10, I demonstrate the application of a saturation plugin to a drum subgroup in a mix. In this case, I apply the saturation plugin to distort the low mids and lows slightly. The application of signal processing across numerous tracks at once, as also heard in Audio Examples 4.3–4.6, satisfies the Gestalt Law of Common Fate. This law states that when objects within an observer's (or listener's) perceptual field are altered to resemble one another, these objects tend to be perceived to belong to a larger group.

At first, the example plays with no processing applied to the drum subgroup. After 10 seconds, I reintroduce the clip with the above saturation settings applied. Listeners will notice that the low frequencies stand out more and also sound more cohesive.

In addition to subgroups, mixers also use auxiliary buses in order process multiple sounds at once. When this technique is used, one can separately control (i) the individual tracks being sent to this auxiliary bus as well as (ii) the auxiliary bus itself. Thus, the mixer can balance the *wet* (i.e., more heavily processed) auxiliary channel and the *dry* (i.e., often less processed) input channels. This technique, where processed and unprocessed versions of the same sound coexist within a given project, is also known as parallel processing.

Auxiliary channels typically feature heavy processing by way of reverb and delay, EQ, and compression tools. However, any effect may be applied to these channels. Once these auxiliary buses are ready to accept input, the mixer can choose how much of the original *dry* signal is present versus the *wet* signal.

Audio Example 4.11

In Audio Example 4.11, I demonstrate the application of delay and reverb via auxiliary sends. This commonly used technique conforms to the Gestalt Law of Common Fate, which states that when multiple objects share similar perceptual characteristics, the observer (or listener) tends to group them together.

In this example, there is a drum groove, bass, pad, and melodic instruments in the form of vocals, synths, and guitar. In order to make these melodic instruments sound more cohesive, I send them all to the same auxiliary channel. This approach causes the melodic instruments to take on identical ambient characteristics, causing listeners to *believe* that these sounds emanate from the same place.

In the above example, the coexistence of both a *wet* and *dry* version of the same signal provides a method for altering a sound while simultaneously keeping most of its recognizable features intact. Thus, the sonic results of this type of parallel processing can be quite subtle, especially in comparison with more direct methods of signal processing. However, when these subtle processing modifications are heard across several tracks at once, the technique's overall perceptual impact is much clearer.

4.7 THE FIGURE-GROUND PHENOMENON: MORE APPLICATIONS FOR MIX BUSES

In the above section, I discuss the use of auxiliary channels and buses for the purpose of processing batches of sounds at once, and I describe how this technique conforms to Gestalt laws of Similarity and Common Fate. Another purpose for using these audio routing features is to provide different processing treatments for *foreground* and *background* tracks within a mix. In other words, mixers can direct the listener's attention by (i) grouping similar sonic elements and (ii) deciding which of those groups will be most prominent on the recording.

In Gestalt theorization, the figure-ground phenomenon describes our tendency to pay attention to one audio stream at a time. Even though we feel like we should be capable of paying attention to several streams at once, this simply is not how we perceive sound [10]. More casually, this phenomenon is also referred to as the "cocktail party effect," or the tendency of our perceptual mechanisms to bring to the foreground whatever conversation we happen to be focused on. For visual artists, the figure-ground phenomenon is often demonstrated via imagery wherein the viewer's perception of foreground and background shifts depending on where they look (Figure 4.3).

Figure 4.3 Demonstrates the figure-ground phenomenon. Although it is a rather simplistic illustration, at first glance viewers may see this image as either four black triangles or a single white diamond.

When producers mix beats, they carefully consider which sounds will comprise foreground elements and which will remain in the background. This is why some producers, such as researcher/engineer Jay Hodgson, describe mixes, in general, as sonic artefacts which have already been heard for us (2014: 36). The mix engineer decides, from their personal aural vantage point, how future listeners will hear the sounds presented within the mix.

While this book doesn't consider rap production in great detail (as this would require another volume), the sonic contrast between vocals and beats in hip-hop music provides the clearest example of the figure-ground phenomenon in action. For instance, hip-hop beats typically support a rap vocal, and, thus, the final mixdown will probably require the mix engineer to highlight the vocal using signal processing and gain staging. This will cause the vocal to stand out in comparison to the instrumental. However, the instrumental may also require some type of opposite treatment in order to draw less attention to it.

In beat-making, producers also choose which sonic elements to highlight or obfuscate within the mix. For instance, many producers divide tracks into percussion, bass, harmony (i.e., strings, pads, guitars, pianos, and similarly rich sonic textures), and effects. In the example below, I experiment with several balances between these groups. In each demonstration, I only modify the gain faders associated with each subgroup.

Audio Example 4.12

In Audio Example 4.12, I demonstrate the figure-ground phenomenon as it applies to mixing. In this audio example, I have drums, bass, pads, and melodic instruments each assigned to a different subgroup. I alter the balance of the mix by simply changing the gain of these subgroups. Listeners will notice that the track sounds very different as I sonically *foreground* each of the different subgroups in this mix.

From 00:00–00:10, the drums are featured more heavily. From 00:10–00:20, the pads are featured more heavily; from 00:20–00:30, the melodic instruments are featured more heavily.

Another potential advantage to mixing tracks in subgroups is that it affords mixers the ability to make small changes to the record's overall balance throughout its duration. For instance, a mixer may choose to use louder drums or bass on the chorus section, whereas the verses may need louder keyboard pads. Approaches to mixing subgroups vary greatly based on the intended aesthetic goal of the producer or mixer.

Audio Example 4.13

In Audio Example 4.13, I demonstrate how a track's balance can change from section to section (i.e., from verse to chorus). In this audio example, as the track transitions into the chorus, I add a number of instrumental layers. These additional layers cause the drums to sound as though they are in the background of the mix. To foreground the drums more, I simply automate the gain on the drum subgroup to increase when the chorus occurs.

4.8 THE RULE OF GOOD CONTINUATION: PARAMETER AUTOMATION AND LOW FREQUENCY OSCILLATORS

The Gestalt *Law of Good Continuation* is also useful for describing the process of mixing records. This law states that changes in a sound's frequency distribution, dynamic characteristics, and location tend to be smooth and continuous and abrupt changes in any of these parameters may cause auditory stream segregation in listeners (Purwins et al. 2008: 158). In fact, the Law of Good Continuation (sometimes referred to as the Law of Good Gestalt) is strongly related to how humans interact with sound in an ecological sense (Moore 2012: 303–313). Of course, as demonstrated in foundational studies by Haas (1972), van Noorden (1975), Bregman and Rudnicky (1975), and others, human hearing is very sensitive to sudden changes in a sound's spectral characteristics, loudness levels, and location of origin. It is widely theorized that the capacity to hear these subtle sonic components directly resulted from evolutionary adaptations of earlier hominids (Moore 2012: 303–313).

One study that demonstrates the Law of Good Continuation was conducted by Darwin and Bethell-Fox (1977: 665–672). They used a synthesizer to create spectral patterns which resembled vowel sounds. For the experimental condition wherein the sound's fundamental frequency (i.e., F0) or overall pitch was constant, listeners perceived the speech-like patterns to form a single auditory stream. However, when F0 moved stepwise to other pitches, the test subjects perceived the vowel sounds to emerge from separate auditory streams. In this later condition, where F0 was not constant, the impression of a good sonic *continuation* was broken, and test subjects perceived the formation of two distinct audio sources rather than one.

When listeners are confronted by sounds which deviate from their sonic expectations, as outlined in Darwin and Bethell-Fox's study, they often perceive these sounds to emerge from separate audio sources. In particular, humans tend to have a number of subconscious expectations for how a sound will transform over time. When we present sounds on records, their spectral transformations, dynamic qualities, and locational qualities throughout the running duration greatly impact the listener's sense of auditory stream formation.

During mixing, producers must decide which sonic elements should transform in a way that sounds continuous and smooth and which elements should transform more abruptly. Either of these techniques can be equally important for establishing the record's overall aesthetic. Consider, for instance, the smooth transition heard when a producer causes a filter to remove low frequencies over the course of several seconds in a track (Audio Example 4.14). When this occurs, the specified frequencies are slowly taken away, yet the listener does not perceive the transformed audio to emanate from a new source. Instead, the listener hears the altered form of the audio to be a *continuation* of the original audio.

Audio Example 4.14

In Audio Example 4.14, I replay the audio from Audio Example 4.13. This time, I ask listeners to pay attention to the filter which is applied to the stereo bus as the track transitions from verse to chorus. In this example, frequencies are removed in a continuous stepwise sequence over several seconds. A momentary silence occurs, and then the full frequency spectrum returns on beat one of the chorus.

This example satisfies the Law of Good Continuation by (i) slowly removing frequency content in a continuous fashion before (ii) returning the lost frequency information. This maneuver can said be a good continuation since the initial filter automation is continuous and smooth. In fact, this sounds very similar to the sonic effect which is produced when one walks away from a loudspeaker, wherein the gain of certain frequency bands is reduced as the listener moves away.

The above example describes a technique known as *parameter automation*, which allows producers to change a track's signal processing strategy throughout its running duration. Today, for most beat makers, automation occurs within a DAW. They automate tracks by either (i) *recording* the changes they make to plugins in real time or (ii) by entering in specific automation changes via a graph or value chart. In the first case, users simply arm their chosen DAW to record plugin parameter changes, and then they modify these parameters while the track plays.

Parameter automation allows producers to change a sound's timbre by enacting one or more parameter modifications throughout the duration of the track. Most of the time, producers tend to automate gain controls and faders, as well as wet/dry mix controls. However, almost any plugin parameter can be automated. For instance, modulation effects such as phasers and chorus effects can sound very interesting when delay times are configured to change as the track plays (Audio Example 4.15).

Audio Example 4.15

In Audio Example 4.15, I demonstrate the automation of delays and reverb in connection with a simple melody. From 00:10–00:20, I slowly increase the wet/dry on the reverb and delay in order to provide a good continuation. From 00:20–00:30, I slowly reduce the wet/dry setting.

While the above example demonstrates a smoother approach to parameter automation, it is also common to use this technique to create abrupt changes in timbre. When this happens, it tends to occur on the first beat of a new section. Consider, for instance, songs which begin with a lo-fi intro, such as "Girls" by Innanet James (2016). After the track plays for several seconds with an absence of low and high frequencies, these frequencies are reintroduced quite abruptly at 00:13, on beat 1 of a new bar.

Indeed, both abrupt and continuous sonic impressions are important for those who mix records. The examples below demonstrate instances where listeners tend to separate auditory streams, as well as instances where listeners tend to perceive cohesive auditory streams.

Audio Example 4.16

In Audio Example 4.16, I demonstrate an abrupt continuation, as well as a good continuation. At first, a vocal melody plays with no additional processing applied. Then, at 00:05, I abruptly change the timbral characteristic in the vocal by boosting the mids by several dB at 1Khz.

At 00:10, I replay this clip with a more common processing strategy which reflects the Law of Good Continuation. First, the clip plays no additional processing applied. Between 00:15–00:20, I gradually add in the frequency boost outlined above.

Audio Example 4.17

In Audio Example 4.17, I demonstrate an abrupt continuation, as well as a good continuation, in terms of a track's dynamic structure. At first, listeners will hear a sample mix with drums, bass, pads, and a melody. Then, at 00:05, I abruptly change the gain on the drums by reducing the gain on this subgroup by several dB. At 00:10, the gain abruptly returns for the chorus section.

At 00:15, I change my approach and slowly reduce the gain of the drums over several beats before returning the lost gain for the chorus section.

Audio Example 4.18

In Audio Example 4.18, I demonstrate an abrupt continuation, as well as a good continuation, in terms of a recording's stereo field characteristics. At first, listeners will hear a sample mix with drums, bass, pads, and a melody. Then, at 00:05, I abruptly change the panning of the melodic instrument by shifting it 100% to the left speaker. At 00:10, I shift this instrument 100% to the right speaker.

At 00:15, I change my approach and slowly move the melodic instrument towards the left (100%) and right (100%) portions of the stereo field in a smooth and continuous manner.

A number of previously released records demonstrated the techniques above in various ways. For instance, in the song "Cartoons and Cereal" by Kendrick Lamar (2012), producers THC and J. LBS cause Kendrick's vocal to pan from the left to the right side of the stereo field once the vocal commences. Engineers and beat makers often automate sonic parameters this way in order to reinforce a song's various compositional sections. For instance, it is common practice to turn up effects sends such as reverb and delay during important sections of the song.

Another common parameter automation technique involves using a low frequency oscillator (LFO). An LFO generates sound waves below 20 Hz, which humans cannot hear, in order to control other sounds or sound-shaping tools. In addition, these LFO rates can be automated to change throughout the running duration of a track. Many music makers and fans associate these devices with synthesizer-based music, like dubstep, and LFOs are integral to the classic dubstep "wub" sound. However, LFOs are also used in conjunction with other effects processors. For instance, in Audio Example 4.19, I apply an LFO to modulate a single band filter in order to cause its cutoff frequency to increase and decrease. The cutoff frequency on the filter simply rises and falls in sympathy as the LFO begins and ends its wave cycle. Sonically, this technique creates periodic recurring changes in frequency distribution, which also conforms to the Gestalt Law of Good Continuation. Indeed, LFO-based parameter modification can enable the smooth and continuous sonic changes described by this perceptual rule.

Audio Example 4.19

In Audio Example 4.19, I use an LFO to achieve a smooth and continuous sonic transformation (i.e., a good continuation). LFOs emit a very slow-moving low frequency energy which can be harnessed to control various FX parameters within a DAW. In this example, I play a Wurlitzer sample with no additional processing (00:00–00:05). After 5 seconds, I apply an EQ filter which is being automated by an LFO. Listeners will notice that the frequencies change smoothly and continuously, and there is an increased sense of sonic *movement* when the filter and LFO are applied.

Producers may also use LFOs in conjunction with stereo panning tools. In Audio Example 4.20, I demonstrate the application of an LFO to a stereo imaging tool. Depending on the settings used, the LFO can cause the stereo panning device to move more quickly or slowly. Faster panning settings emulate the projection of natural sounds, which tend to reach our ears at different points in time. Interestingly, slower panning settings do not necessarily sound less natural than faster settings, as the LFO creates a smooth and continuous transformation over time.

Audio Example 4.20

In Audio Example 4.20, I use an LFO to achieve a smooth and continuous sonic transformation (i.e., a good continuation) as the audio moves from the left to the right speaker. LFOs emit a very slow-moving low frequency energy which can be harnessed to control various FX parameters within a DAW. In this example, I play a Wurlitzer sample with no additional processing (00:00–00:05). After 5 seconds, I apply a panning plugin which is controlled by an LFO. Listeners will notice that the track's stereo configuration changes smoothly and continuously as the audio moves from the left to the right speaker and back.

4.9 CHAPTER SUMMARY

In order for listeners to accept a recording as a musical document, the document must be configured so that they accept it as one. In a similar way, a book might contain a novel, reference information, phone numbers for every person in your city, a collection of maps, a number of essays, or stale jokes from 30 years ago. No matter what the book contains, a number of preconditions must be satisfied in order for the reader to recognize its purpose.

In record production, when these preconditions are satisfied, the record instills a sense of *sonic realism* within listeners—a form of aesthetic realism which permeates all types of commercial record production. Different genres simply present different flavours of sonic realism. For instance, 1970s roots rock, recordings of gamelan music, and trap all adhere to this broad aesthetic sensibility while at the same time sounding completely different from one another.

The sonic differences which cause listeners to distinguish these types of communications are not insignificant. For example, it takes a very different type of approach to create a hip-hop beat versus an indie folk record. In order to create either document, the producer has to decide at the outset what kind of project s/he would prefer to create. Once this decision is made, there are many ways in which a producer can configure musical materials to conform to the general aesthetic guidelines they have chosen. As I've shown in Chapters 2 and 3, many aesthetic qualities in record production are directly measurable, such as a record's dynamic configuration, spectral design, and stereo correlation.

Having discussed CBMP platforms, CBMP musical materials (i.e., samples, synths, synth presets, and mix tools), mixing, and the psychophysical dimensions of hip-hop record production, I now turn the reader's attention to a number of case studies in the next chapter.

NOTES

1. Some projects will rely on a group consensus of some type in order to decide on a record's final sonic characteristics. However, it is almost always a single engineer who will interpret the requests and opinions of the other involved parties since it is impractical for multiple people to work on the same project, at the same time, in the same place.
2. Moylan (2014), Hodgson (2010, 2017), Harding (2017), Paterson (2017), Izhaki (2008), Krotz (2017), Shelvock (2011, 2012, 2014, 2016, 2017b), and Nelson (2017).
3. An in-depth discussion of what music is versus what it is not; a broad discussion of the musical genre far exceeds the scope of this book.
4. Historically, many recording tasks are performed separately. The traditional way of approaching record creation is to (i) track and sequence sounds, (ii) edit these sounds, (iii) mix these sounds, and then (iv) master them. However, recordings rarely proceed exactly this way, and recordists often freely switch between tasks (i)–(iii) in order to finalize a track's mix. For instance, one might notice that a bass synth and kick drum blend to create a nice timbre when the bass transient occurs milliseconds after the kick drum transient. A mix engineer might edit each instance where the bass instrument occurs in order to create a different timbral aesthetic. Such a maneuver impacts the mix in an obviously audible way, but editing is often thought to be a separate task from mixing. I argue that in beat-making in particular, distinctions between sequencing, editing, and mixing are far less meaningful than other production genres.

5

Case Studies in CBMP

This chapter provides listeners with case studies in cloud-based music making. In each section, I provide an explanation of how each track was created using CBMP tools, as well as audio examples which clarify this process further. Each of the below case studies is based on a record which is available on commercial DSPs such as Spotify, Apple Music, and others. Readers are encouraged to follow along with the audio examples which are listed in the text below.

CASE STUDY 1: "MELANCHOLY" BY KINGMOBB (2019)

Overview

This record, "melancholy" (2019), elicits a sad and contemplative feeling (Audio Example 5.1). It is a slow track (66.80 bpm) in comparison with the average hip-hop record. However, some popular tracks, such as "O.P.P." by Naughty by Nature (1999; 65 bpm), "Heard 'Em Say" ft. Adam Levine by Kanye West (2005; 63 bpm), and "Light Up" by Drake (2010; 67 bpm) are also in this tempo range.

Of all of the case studies in this book, the production process for this track was the simplest. In fact, this record is made from 5 continuous samples, all of which I acquired using a CBMP platform. I invite readers to re-create this case study by incorporating the following samples into their favourite DAW (while ensuring that the time-stretch feature is on):

1. BBL_75_vocal_best_friends_Am.wav (Capsun Audio)
2. BBL_75_drum_loop_mellow_muffle.wav Capsun Audio)
3. RF_126_A_ACOUSTIC_BASS.wav (Live House Elements by Loopmasters)
4. BBL_90_textured_melodic_clockwork_Am.wav (Capsun Audio)
5. RA_UKHH_90BPM_Vinyl-Hiss-03.wav (Rankin Audio)

Listeners will notice that the track's timbral aesthetic is rather vintage and *lo-fi*. Lo-fi hip-hop is a subgenre wherein producers intentionally use samples which sounds as though they are being played through faulty analog equipment of some type. As a result, records in this genre simultaneously feature hiss and crackle, as well as warm-sounding low-midrange frequencies which feature pleasant sounding harmonic distortion. I have re-created this aesthetic in the current track.

Each of the samples used on this record have been time stretched. This causes each sample's timbral characteristics to be slightly altered. I demonstrate the difference between the time-stretched and non-time-stretched version of the bass in Audio Example 5.2. At first, listeners hear the original sample at 63 bpm. Then, I replay the sample at 66. Listeners will notice that the faster sample sounds punchier and more present. Even though the tempo change is only small, it results in a slightly different timbral characteristic. This happens because when the track is sped up, a greater number of transients occur in a shorter period of time.

Constructing the Main Theme

For myself, and many other producers, the most important step in beat-making involves the creation of a 4–8 bar thematic segment. In the very least, these segments typically feature drums, bass, and some assortment of atmospheric and textural samples. However, these segments can also be much more involved, featuring melodies, counter melodies, harmonies, FX, and other sounds. In the production world, these short segments are referred to as *loops*, even though, in many cases, they are also made from loops (or a combination of loops, custom passages made from one-shots, and perhaps other sounds).

For "melancholy," I began with an 8-bar segment, based on the vocal sample called BBL_75_vocal_best_friends_Am.wav (Capsun Audio) and a drum sample called BBL_75_drum_loop_mellow_muffle.wav (Capsun Audio). The vocal sample was tuned down by two semitones. From this point, I added the bass sample. I have included two variations of this sample. The first type lasts for 1 bar, and I have tuned it down by 6 semitones. The second variant lasts for 1 bar, and I have tuned it down by only 2 semitones. In terms of musical arrangement, this makes the track sound as though it is alternating between chords I and iii.

From this point, I added a pad sample called BBL_90_textured_melodic_clockwork_Am.wav (Capsun Audio) and a vinyl hiss sample called RA_UKHH_90BPM_Vinyl-Hiss-03.wav (Rankin Audio, Figure 5.2). These two samples strongly reinforce the track's lo-fi aesthetic. Perhaps even more than the drums and vocals, which also exhibit a lo-fi characteristic, the pad and hiss samples give the listener the sense that s/he is listening to some type of vintage audio equipment.

When producers construct the main theme for a track, they often perform some mixing. Before working on this track's form or arrangement any further, I added a mid boost of several decibels to the vocal and drum samples (i.e., around 1Khz with a wide Q-value). I also boost the low frequencies in the bass by several decibels (ie., around 80 Hz). In terms of ambient processing, this track has one reverb send, armed with a simple medium room reverb (i.e., Ableton Native Plugin) panned 25% to the left. I also added sidechain compression to the pads, bass, vinyl hiss, and reverb send channel. This sidechain compression is triggered by the kick drum, as heard in Audio Example 5.3.

Sketching Form

Once I was satisfied with the above loop, I began to sketch out the track's overall form. In this case, I decided I want the record to begin with a short fade in. During this time, only the pad sample is playing. After 7 seconds, the track's main theme plays. I copy and pasted the track's main theme four times, at which point the track fades out to the sound of the pad sample.

In each of the four repetitions of the main theme, I use editing in order to (i) alter the timing of the vocal line, (iii) mute instruments, and (iii) create rhythmic variations in the drums. I demonstrate these variations in Figure 5.1.

Final Mix and Master

Hip-hop beat-making has a strong DIY tradition, and, as a result, many beats in this genre are not professional mixed and mastered. This track, "melancholy," is no exception. In this case, most of the mixing was finished when I completed the track's main theme. I simply mastered the track by applying Waves Center and the Fabfilter Pro L to the master bus.

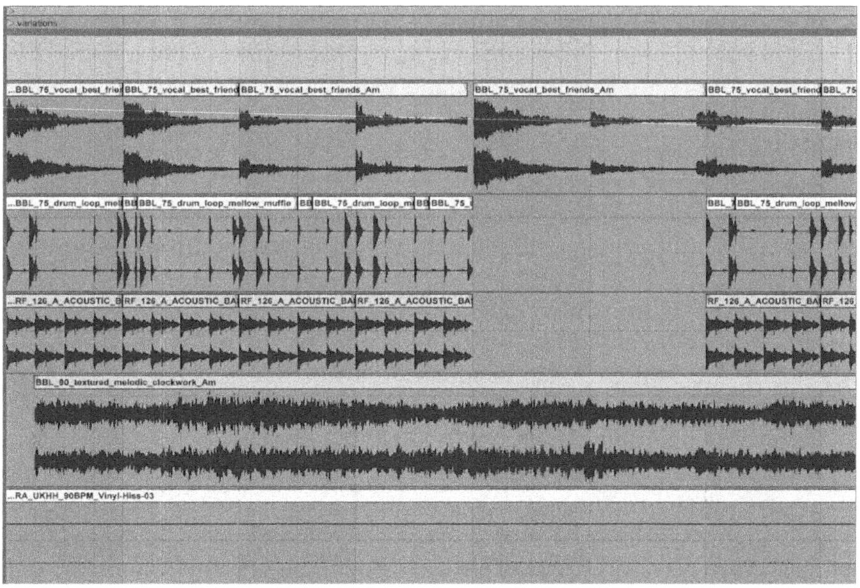

Figure 5.1 Provides a screenshot which demonstrates how I alter the drums, vocals, and pads in order to create variation within the arrangement. As seen in this photo, I simply cut and paste various parts and occasionally mute them.

In Audio Example 5.4, I demonstrate the track with and without this signal chain applied.

CASE STUDY 2: "FAYTH" BY KINGMOBB (2019)

Overview

This case study explains how I created "fayth" by kingmobb (2019, Audio Example 5.5). This record features both vintage and modern timbres in the form of continuous samples, one-shot samples, and a bass synth. Each of these sounds, including the bass synth patch, was acquired using a CBMP platform.

Samples:

1. 80_BoomBapBeats_3_TL.wav (Touch Loops)
2. fs_insp90_chill_main_vox2_Am.wav (Sample Magic)
3. CPA_SPK_130_4_Bar_Loop_03_Gm_Vintage.wav (Capsun Audio)
4. HDVA_Hugetexture_Dmin_128bpm.wav (Black Octopus)

Synth (NI Massive):

1. BS H8OhH8.nmsv (Essential Trap Presets for Massive by Primeloops)

Constructing the Main Theme

In order to construct this track's main thematic loop, I began with a drum sample called 80_BoomBapBeats_3_TL.wav (Touch Loops), which I sped up to 85 bpm. I regularly curate my own sample library, so I had a number of melodic samples on hand which could fit this piece. I opted to use fs_insp90_chill_main_vox2_Am.wav (Sample Magic) since the track's bpm is very close to this melodic sample's original bpm (i.e., 90 versus 85 bpm).

The melodic sample contains a haunting vocal, which I transposed down by 7 semitones. In addition to lowering the pitch, this alteration significantly transformed the timbre of the sample. The downtuned version sounds boxier with an abundance of low-mids. There is also an audible *wooshing* sound, similar to a phaser or a flanger, present in the re-pitched version of the sample. I demonstrate the difference between the original sample and the version from "fayth" in Audio Example 5.6.

Once the vocal and drums were in place, I added a bass synthesizer to the arrangement. More specifically, I used Massive, a wavetable synth made by Native Instruments. I downloaded a patch from Splice called BS H8OhH8.nmsv (Essential Trap Presets for Massive by Primeloops) and proceeded to play various bass lines using a MIDI controller until I settled on G0 (2 bars), D#0 (2 bars), C0 (2 bars), and G0 (2 bars).

From this point, I added keyboards by (i) finding a Wurlitzer sample on splice called CPA_SPK_130_4_Bar_Loop_03_Gm_Vintage.wav (Capsun Audio),

(ii) tuning this sample to the key of this track, (iii) chopping the sample into segments which last 2 beats, and (iv) replaying these samples using a MIDI controller (Audio Example 5.7 plays this track in isolation). In Figure 5.2, I provide a screenshot of the MIDI data I recorded for this track, as well as the corresponding audio data.

Once this keyboard part was complete, I decided to add some additional pads in order to provide some textural support for the track's main elements. The pad I chose is called HDVA_Hugetexture_Dmin_128bpm.wav (Black Octopus).

Sketching Form

Like many hip-hop instrumentals, this beat has a very simple form. In this case, the use of an unsophisticated structure can allow emcees to easily load this track into a DAW and record with it. This also allows them to loop the performance as many times as they would like. After 1:10, I decide that the main vocal sample has become fatiguing to listen to, and I mute it. At the same time, I automate the volume of the pad to increase in volume. This pad sounds as though it contains a French horn, and the effect creates a sort of cinematic sounding bridge three-quarters of the way through the track.

As with Case Study 1, I create variation within this record's arrangement by simply editing and changing the order of segments of the drum and vocal samples.

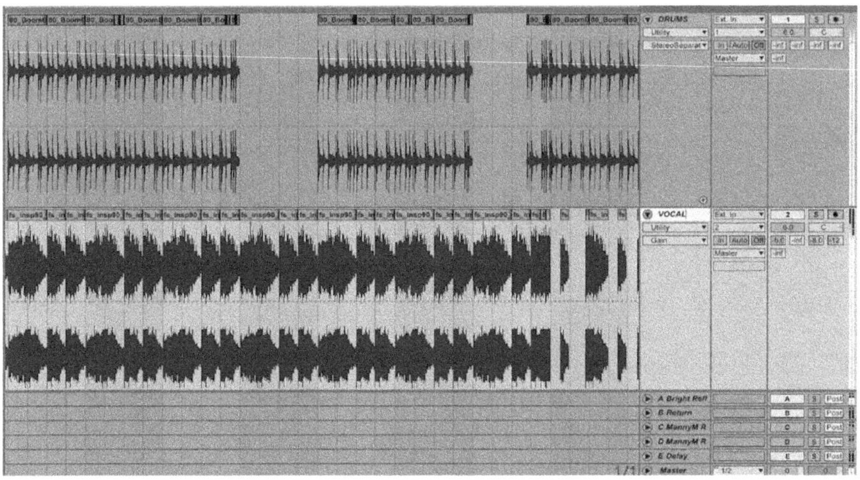

Figure 5.2 Provides a screenshot which demonstrates how I alter the drums and vocals in order to create variation within the arrangement. As seen in this photo, I simply cut, paste, and rearrange various parts and occasionally mute them.

Final Mix and Master

Since this track also features a number of dense samples, the mixing process was fairly simple. In fact, when mixing these types of *complete sounding* samples, the process resembles the practice of mixing using subgroups (as outlined in Chapter 4). Since all of today's CBMP audio samples are created by professional audio engineers and producers, they often require very little correction or alteration unless the producer wishes to do so.

However, one mixing technique which is used consistently across many sample-based instrumentals is sidechain compression. In this example, I have place compressors on the Wurlitzer, pad, and bass tracks which cause these instruments to attenuate each time the kick drum sounds. By also applying the sidechain EQ filter on each of these compressors, I can cause the compressor to ignore all incoming sounds above 80Hz.

To master this track, I have, once again, used Waves Center and the Fabfilter Pro L limiter on the master bus. I demonstrate the unmastered versus mastered version of this track in Audio Example 5.8.

CASE STUDY 3: "DXN'T WANT U" BY KINGMOBB, FROM *KNGMXBB II* (2017)

In this case study, I cover the production process for the track "Dxn't Want U," which was also released under my moniker, kingmobb, from *KNGMXBB II* (2017, Audio Example 5.9). Unlike the previous two case studies, this track has a more modern timbral characteristic. In addition, this track make more extensive use of one-shots and synthesizers than the other two case studies. Here are the samples and synthesizer presets I used to create this track:

Vocal:

1. (Melodic) BS2_80_G_IDontWantY.ouNoMore_2.wav (Loopmasters)
2. (Spoken) SR_140_VOCAL_LOOP3_FX.wav (Soul Rush Records)

Melody:

1. SCT_Melody_Loop_27_F#_130_Bpm.wav (Singomakers)
2. DLTT_Arpeggio_06_100_Cm.wav (Capsun)

Drums (one-shots):

1. FM_kick_trap.wav (Splice)
2. Just_snare_ringer.wav (Splice)
3. KRNE_Movin_Snr.wav (Splice)
4. Perc_Above2.wav (Julez Jadon)
5. Perc_Above1.wav (Julez Jadon)

Additional percussion (continuous samples):

1. PercLoop_Abacus_95BPM.wav (Julez Jadon)
2. ShakerLoop_Temple_85BPM.wav (Julez Jadon)

FX:

1. LL_riser_pink.wav (Splice)
2. JYKD_fx_impact_hit.wav (Splice)
3. Wafx_impact_guncrime2.wav (Wave Alchemy)
4. NT2_GUNSHOT.wav (Loopmasters)
5. Gun_Cock.wav (Loopmasters)

NI Massive bass:

1. BS H8OhH8.nmsv (Essential Trap Presets for Massive by Primeloops)

NI Massive pads (stock):

1. Harmonia.nmsv
2. WavinEighties.nmsv

NI Massive synths (stock):

1. Sebasstian.nmsv
2. StabMau5Pluck.nmsv

Constructing the Main Theme

The main loop in "Dxn't Want U" is based on a vocal sample called BS2_80_G_IDontWantY.ouNoMore_2.wav (Loopmasters). After I imported this sample, I increased the tempo of this track to 105 bpm, and then I began audition various drum sounds. One I had settled on FM_kick_trap.wav (Splice) for the kick drum and the combination of Just_snare_ringer.wav (Splice) and KRNE_Movin_Snr.wav (Splice) for the track's primary snare drum, I recorded myself triggering these samples with a MIDI controller. After some editing, I finalized the track's main drum part (Audio Example 5.10).

From this point, I added an instance of NI's Massive, a synth plugin, and I loaded a preset I downloaded from a CBMP service called BS H8OhH8.nmsv (Essential Trap Presets for Massive by Primeloops). I controlled this synth using a MIDI controller and recorded a number of different ideas until I decided upon the track's bass line (Audio Example 5.11).

Drawing inspiration from musicians like So Loki, Knxwledge, Kaytranada, Mick Jenkins, Mouthe, and others, I decided that this track should have a number of auxiliary percussion sounds rather than a conventional hi hat. I repeated this strategy on the song "Cut Deep" from the same album (2017). In this track, I used 2 continuous percussion samples called PercLoop_Abacus_95BPM.wav and ShakerLoop_Temple_85BPM.wav (Julez Jadon) (Audio Example 5.12).

Once the above elements were in place, I decide that I wanted to find a melody to work as a chorus, or B-section, for this record. I found a melodic sample called DLTT_Arpeggio_06_100_Cm.wav (Capsun) for this purpose, which becomes audible about 45 seconds into the track (please see Audio Example 5.9 again). This sample was obfuscated by some other sounds in the mix, so I opted to layer this sample with two synthesizers which play the same melody. By reinforcing the melody with similar elements, I also demonstrate the Gestalt Law of Similarity covered in Chapter 4. Listeners perceive the additional synthesizers and the original sample to exist as a single auditory stream since they have similar rhythmic and pitch characteristics. In Audio Example 5.13, I play the original sample for several seconds, and then it plays once again with the additional synthesizers added. Listeners will notice the reinforcing effect in the latter half of the example.

Sketching Form

Once I had settled on an A melody (using BS2_80_G_IDontWantY.ouNoMore_2.wav, by Loopmasters) and a B melody (using DLTT_Arpeggio_06_100_Cm.wav by Capsun), I simply copy and pasted these segments across the project in an alternating format with drums and bass (i.e., ABABABA).

To add interest, at 1:21, I add an emcee sample called SR_140_VOCAL_LOOP3_FX.wav (Soul Rush Records). I layer this vocal with the track's B melody (please see Audio Example 5.9 again). Listeners will notice that, in this section, I use a number of samples to liven up the arrangement. These include LL_riser_pink.wav (Splice), JYKD_fx_impact_hit.wav (Splice), Wafx_impact_guncrime2.wav (Wave Alchemy), NT2_GUNSHOT.wav (Loopmasters), and Gun_Cock.wav (Loopmasters). In Audio Example 5.14, I demonstrate these FX in isolation.

Final Mix Versus Master

Unlike the other two case studies, the record discussed in Case Study 3 was mastered professionally by Jedi Mastering in Ontario, Canada. In Audio Example 5.15, I present a short snippet of the final mix (15 seconds), as well as a segment of the master copy. I do this so that listeners can compare the original version with the final result.

Conclusions

In this book, I've focused on key CBMP features which shape the way that many producers—at least 2.5 million active Splice users (and many more who use alternative platforms such as Loopcloud, Noiiz, 99 sounds, and others)—create records (King 2019). I have taken readers through numerous applications for CBMP tools within the practice of creating hip-hop beats because this genre features a number of distinctive music-making practices which CBMP services overtly support. Readers should note that numerous additional applications for these tools exist; what I have covered in this book is by no means exhaustive. However, I have sought to ground my analysis of this new creative paradigm within a discussion of some tangible production skills by pairing the discussion of CBMP features with bona fide hip-hop studio practices.

Now that the means to create and disseminate music are so easily accessible—requiring only a phone, tablet, or computer—new types of musical skills and literacies are now emerging at a more rapid pace than ever before. However, at the same time, authors only just beginning to substantively comment upon studio operations from the point of view of those who make records. Until recently, the scholarly conversation on music production has been dominated by record collectors, aficionados, theorists, and journalists who, in the absence of more accurate data, were once seen as the ultimate authorities on the subject.

In seeking to directly comment upon the act of record creation, this book is different from the literature I describe above. For instance, I have aimed to move past the (often phenomenological) post hoc analysis of available records, which has been the dominant paradigm for discussing record creation. There is certainly nothing wrong with such analyses, and, on occasion, this book also provides some similar post hoc analyses of available records as evidence for the popularity of the studio techniques I describe. However, I also provide a *direct* commentary upon the techniques which create these sounds, as opposed to simply identifying or cataloguing them. In this book, I have (i) provided readers with a general digital taxonomy of CBMP audio samples and synthesizer presets (which explains the sonic categories which producers actually use within searchable databases), (ii) explained how these samples are used to create music, (iii) elucidated how these samples and synthesizer presets are altered to fit together within a mix composite, (iv) explained the perceptual impacts of these so-called mixing techniques, and (v) provided case studies which provide context for the production operations discussed throughout this book's Introduction and Chapters 1–4. Thus, as a unique feature of this book, I have maintained a direct focus on the creative inputs which *create hip-hop beats*, rather than the identification of these same techniques *on available records*, like so many other papers and books have already done.

To be sure, such *post hoc* analyses of records can be useful and insightful. However, there is a tendency in this type of literature to confuse the *inputs* which *create* records for the aural identification of these same techniques on available records. And, unfortunately, many authors who do so adopt abstracts and titles which seem as though they are discussing music production, yet they only address the *result of* music production efforts. It is my belief that this approach is confusing, and, in some cases, it can create hurdles for those whose primary research activities deal with music production *per se*.

We have recently begun to hear more from researchers whose primary focus is on record creation. The field of Music Production Studies (MPS), which was recently announced (c. 2017) by authors Russ Hepworth-Sawyer, Jay Hodgson, and Mark Marrington, has significantly advanced research which examines music production in a more direct way. This book is aligned with the goals of MPS. Given the popularity of recent research (c. 1996–2012) which examines the cultural and theoretical dimensions of music production and record consumption, the next reasonable step is for authors to build a more accurate body of literature which covers the act of record creation *per se*. This type of literature is beginning to gain momentum within scholarly circles, but, because MPS is an emerging field, there are still numerous topics which remain uncovered.

By focusing on the use of CBMP, I hope to have clarified one such gap in music production literature. Maintaining a focus on CBMP has also enabled me to shed some light on the under-reported practice of *lawful* sampling, as well as methods for creating music from these samples. In addition, I have also clarified some of the broad categories producers now use to speak about sounds—whether they come from synthesizer presets or audio samples. As I argued in this book's Introduction, aurally identifying the use of audio samples or synth presets on hip-hop records, for instance, is quite different from a discussion of how a sample or synth preset was (i) captured, (ii) edited, (iii) sonically altered (i.e., mixed), and (iv) repurposed within a new composition or arrangement. Of course, as practitioners know, each of these steps is rife with sophisticated aesthetic nuances, and the ways that one addresses these nuances *always* determines the end result of a finalized beat (even if it somehow contains truly aleatoric components). Until researchers establish a more thorough account of these studio techniques (in ways which actually reflects their usage within music-making communities), we will not be able to substantively or accurately critique them beyond the act of simply identifying them. What's more, fields of music research might never fully grasp music production practice and its importance in shaping contemporary musical experiences.

If fields of musical analysis fail to evolve their collective methods, bracketing the perspectives of those who create records in favour of older models of theorization (which do not incorporate an understanding of record making *per se*), then our field will not be able to substantively comment upon the way that music is most commonly experienced: recordings. It is my hope that this book, as well as others within this series and the field of MPS, helps to provide a more solid foundation for future investigations of record practice, in a broad sense, as well as topics related to sampling, synthesis, mixing, and beat creation in hip-hop.

Bibliography

Bregman, A. S. (1994). *Auditory Scene Analysis: The Perceptual Organization of Sound*. Cambridge: MIT Press.

Bregman, A. S., & Campbell, J. (1971). Primary Auditory Stream Segregation and Perception of Order in Rapid Sequences of Tones. *Journal of Experimental Psychology: Human Perception and Performance*, *9*(2), 244.

Bregman, A. S., & Rudnicky, A. I. (1975). Auditory Segregation: Stream or Streams? *Journal of Experimental Psychology: Human Perception and Performance*, *1*(3), 263.

Camp, Z. (2015). Just Blaze Releases Early '00s Drum Samples, Making Good on Meow the Jewels Promise. *Pitchfork* [online]. Available at: http://pitchfork.com/news/62149-just-blaze-releases-early-00s-drum-samples-making-good-on-meow-the-jewels-promise/ [Accessed: 18 January 2016].

Caulfield, K. (2019). 30% of All U.S. Music Streams in 2018 Were R&B/Hip-Hop, with Drake the Most-Streamed Artist of Year. *Billboard* [online]. Available at: www.billboard.com/articles/columns/chart-beat/8493091/drake-2018-most-streamed-artist-rb-hip-hop

Cousins, M., & Hepworth-Sawyer, R. (2013). *Practical Mastering*. New York: Routledge/Focal Press.

Danielsen, A. (Ed.). (2010). *Musical Rhythm in the Age of Digital Reproduction*. Farnham: Ashgate Publishing.

Darwin, C. J., & Bethell-Fox, C. E. (1977). Pitch Continuity and Speech Source Attribution. *Journal of Experimental Psychology: Human Perception and Performance*, *3*(4), 665.

Deahl, D. (2017). Splice Is Making It Much Easier to Buy Expensive Music Production Plugins. *The Verge* [online]. Available at: www.theverge.com/tech/2017/10/6/16431394/splice-dj-plugins-ozone-8-neutron-2

Devine, A., & Hodgson, J. (2017). Mixing in/and Modern Electronic Music Production. *Perspectives on Music Production: Mixing Music*. Eds. Hepworth-Sawyer, R., & Hodgson, J. New York: Routledge/Focal Press.

Eisengrein, D. (2007). Do the Dip: Sidechains. *Remix Magazine*. Available at: www.remixmag.com

Gelfand, S. A. (2017). *Hearing: An Introduction to Psychological and Physiological Acoustics*. Boca Raton: CRC Press.

Giannoulis, D., Massberg, M., & Reiss, J. D. (2013). Parameter Automation in a Dynamic Range Compressor. *Journal of the Audio Engineering Society*, *61*(10), 716–726.

Gibson, B. (2002). *Sound Advice on Mixing*. ArtistPro.

———. (2004). *Sound Advice on Equalizers, Reverbs and Delays*. ArtistPro.

The Guardian. (2016). Kanye West: Secret Pirate Bay User? *The Guardian* [online]. Available at: www.theguardian.com/technology/2016/mar/02/kanye-west-secret-pirate-bay-user-serum

Haas, H. (1972). The Influence of a Single Echo on the Audibility of Speech. *Journal of the Audio Engineering Society*, *20*(2), 146–159.

Harding, P. (2017). Top Down Mixing—A 12 Step Mixing Program. *Perspectives on Music Production: Mixing Music*. Eds. Hepworth-Sawyer, R., & Hodgson, J. New York: Routledge/Focal Press.

Hepworth-Sawyer, R., & Golding, C. (2011). *What Is Music Production? A Producer's Guide: the Role, the People, the Process*. New York: Routledge/Focal Press.

Hepworth-Sawyer, R., & Hodgson, J. (2017). *Perspectives on Music Production: Mixing Music*. New York: Routledge/Focal Press.

———. (2019). *Audio Mastering: The Artists*. New York: Routledge/Focal Press.

Hereford, J., & Winn, W. (1994). Non-Speech Sound in Human-Computer Interaction: A Review and Design Guidelines. *Journal of Educational Computing Research, 11*(3), 211–233.

Hodgson, J. (2010). *Understanding Records*. London: Continuum.

———. (2011). Lateral Dynamics Processing in Experimental Hip-hop: Flying Lotus, Madlib, Oh No, J-Dilla and Prefuse 73. *Journal on the Art of Record Production* [online]. Available at: http://arpjournal.com/lateral-dynamics-processing-in-experimental-hip-hop- flying-lotus-madlib-oh-no-j-dilla-and-prefuse-73/ [Accessed: 18 January 2016].

———. (2014). *Representing Sounds*. Waterloo: WLA Press.

———. (2017). Mix as Auditory Response. *Perspectives on Music Production: Mixing Music*. Eds. Hepworth-Sawyer, R., & Hodgson, J. New York: Routledge/Focal Press.

Houghton, E. (2017). Modern Approaches, Hip-Hop: Drums & Percussion. *Red Bull Academy* [online]. Available at: http://daily.redbullmusicacademy.com/2017/10/modern-approaches-hip-hop-drums

Izhaki, R. (2013). *Mixing Audio: Concepts, Practices and Tools*. New York: Routledge/Focal Press.

iZotope.com. (2013). iZotope Interview: "How to Add Tape Saturation to Your Mastering Chain." Available at: www.izotope.com/en/learn/products/ozone/how-to-add-tape-saturation-to-your-mastering-chain-with-ozone-7-advanced.html

———. (2016). iZotope Interview: Just Blaze [online]. Available at: www.iZotope.com/en/community/artists/engineers/just-blaze [Accessed: 18 January 2016].

Katz, B. (2007). *The Art and Science of Audio Mastering*. New York: Routledge/Focal Press.

Katz, M. (2010a). *Capturing Sound: How Technology Has Changed Music* (Vol. 1). Berkeley: University of California Press.

———. (2010b). *Groove Music: The Art and Culture of the Hip-Hop DJ*. Oxford: Oxford University Press.

King, A. (2019). Music Creation Platform Splice Raises a Whopping $57.5 Million in Series C Funding. *Digital Music News* [online]. Available at: www.digitalmusicnews.com/2019/03/20/splice-series-c-funding/

Kohler, W. (1938). Physical Gestalten. *A Sourcebook of Gestalt Psychology*. Ed. Ellis, W. D. London: Routledge & Kegan Paul. (Originally published in 1920.)

Krotz, A. (2017). Mixing for Markets. *Perspectives on Music Production: Mixing Music*. Eds. Hepworth-Sawyer, R., & Hodgson, J. New York: Routledge/Focal Press.

Lehar, S. (1999). Gestalt Isomorphism and the Quantification of Spatial Perception. *Gestalt Theory, 21*, 122–139.

Levine, M. (2018). How to Mix with Distortion, Overdrive, and Saturation. *Waves* [online]. Available at: www.waves.com/how-to-mix-with-distortion-vs-overdrive-vs-saturation

Marrington, M. (2017). Composing with the Digital Audio Workstation. *The Singer-Songwriter Handbook*. Eds. Williams, J., & Williams, K. Bloomsbury.

Miller, G. A., & Heise, G. A. (1950). The Trill Threshold. *The Journal of the Acoustical Society of America, 22*(5), 637–638.

Moore, B. C. (2012). *An Introduction to the Psychology of Hearing*. Leiden, Netherlands: Brill Publishing.

Moylan, W. (2014). *Understanding and Crafting the Mix: The Art of Recording.* New York: Routledge/Focal Press.

Nelson, D. (2017). Between the Speakers: Discussions on Mixing. *Perspectives on Music Production: Mixing Music.* Eds. Hepworth-Sawyer, R., & Hodgson, J. New York: Routledge/Focal Press.

Owsinski, B. (2008). *The Mastering Engineer's Handbook: The Audio Mastering Handbook.* Boston: Cengage Learning.

Pangburn, D. J. (2016). Splice Is Building a Collaborative Tool for Music Production. *The Verge* [online]. Available at: www.theverge.com/2016/4/1/11349478/splice-beatmaker-samples-sequencing

Patterson, J. (2017). Mixing in the Box. *Perspectives on Music Production: Mixing Music.* Eds. Hepworth-Sawyer, R., & Hodgson, J. New York: Routledge/Focal Press.

Pensado, D. (2014). Noah '40' Shebib—Pensado's Place #151 [online]. Available at: www.youtube.com/watch?v=ESUHhXglaos&index=19&list=PLh0d_FQ4KrAVF5LM-2sCC_nncdjk7jCK7G [Accessed: 3 October 2017].

Purwins, H., Herrera, P., Grachten, M., Hazan, A., Marxer, R., & Serra, X. (2008). Computational Models of Music Perception and Cognition I: The Perceptual and Cognitive Processing Chain. *Physics of Life Reviews*, *5*(3), 151–168.

Resnikoff, P. (2019). Hip-Hop + R&B Account for 1 Out of Every 3 Music Streams in America. *Digital Music News* [online]. Available at: www.digitalmusicnews.com/2018/07/09/hip-hop-rap-music-streaming-2018/

Robjohns, H. (2010). Analog Warmth. *Sound on Sound* [online]. Available at: www.soundonsound.com

———. (2014). What Does a Compressor's Auto-Release Do? *Sound on Sound* [online]. Available at: www.soundonsound.com

Ronan, M., Nicholas, W., & Robert, S. (2017). Exploring the Sound Quality Dimensions of Hypercompression. *Innovation in Music II.* Eds. Patterson, J. L., Toulson, R., Hepworth-Sawyer, R., & Hodgson, J. Future Technology Press, 39.

Russonello, G. (2013). Why J Dilla May Be Jazz's Latest Great Innovator. *NPR* [online]. Available at: www.npr.org/sections/ablogsupreme/2013/02/07/171349007/why-j-dilla-may-be-jazzs-latest-great-innovator

Schloss, J. G. (2014). *Making Beats: The Art of Sample-Based Hip-Hop (Music Culture).* Middletown: Wesleyan University Press.

Senior, M. (2011). How to Build Tracks Around Sampled Tunes. *Soundonsound.com* [online]. Available at: www.soundonsound.com/sos/dec11/articles/steal-the-feel.htm [Accessed: 18 January 2016].

Sewell, A. (2013). *A Typology of Sampling in Hip-Hop* (Doctoral dissertation, I).

———. (2014). Paul's Boutique and Fear of a Black Planet: Digital Sampling and Musical Style in Hip-hop. *Journal of the Society for American Music*, *8*(1), 28–48.

Shelvock, M. (2011). Interview with Josh Leo. *Journal on the Art of Record Production*, (5) [online].

———. (2012a). Interview with Steve Marcantonio. *Journal on the Art of Record Production*, (6) [online].

———. (2012b). Interview with Ben Fowler. *Journal on the Art of Record Production*, (6).

———. (2014). The Progressive Heavy Metal Guitarist's Toolkit: Analog and Digital Strategies. *Innovation in Music.* Shoreham-by-Sea: Future Technology Press.

———. (2015). Gestalt Theory and Mixing Audio. *Innovation in Music Conference.* Anglia-Ruskin University, Cambridge, UK. Conference Paper.

———. (2016). Gestalt Theory and Mixing Audio. *Innovation in Music II.* Eds. Patterson, J., Toulson, R., Hepworth-Sawyer, R., & Hodgson, J. Shoreham-by-Sea: Future Technology Press.

———. (2017a). Groove and the Grid: Mixing Contemporary Hip-hop. *Perspectives on Music Production: Mixing Music*. Eds. Hepworth-Sawyer, R., & Hodgson, J. New York: Routledge/Focal Press.

———. (2017b). *Audio Mastering as a Musical Competency* (Dissertation). University of Western Ontario. Available at: http://ir.lib.uwo.ca/etd/

———. (2019). Socio-Technical Networks and Music Production Competencies: Towards an Actor Network Theory of Hip-hop Beat-Making. *Perspectives on Music Production: Producing Music*. Eds. Hepworth-Sawyer, R., & Hodgson, J. New York: Routledge/Focal Press.

Simons, D. (2011). Mark Neill: Recording the Black Keys at Muscle Shoals. *Sound on Sound* [online]. Available at: www.soundonsound.com/people/mark-neill-recording-black-keys-muscle-shoals

Speranza, G. (1979). Looka Here! It's Sam & Dave. *Time Barrier Express*, 26, 3, no. 6. NYC [online]. Excerpts available at: https://quoteinvestigator.com/tag/martin-mull/

Tenney, J., Ashley, R., & Nunn, C. M. (1977). *META Meta+ Hodos*. Frog Peak Music.

Tingen, P. (2007). DJ Premier. *Soundonsound.com* [online]. Available at: www.soundonsound.com/sos/jul07/articles/djpremier.htm [Accessed: 18 January 2016].

———. (2009). Secrets of the Mix Engineers: Young Guru. *Soundonsound.com* [online]. Available at: www.soundonsound.com/people/secrets-mix-engineers-young-guru

van Noorden, L. P. A. S. (1975). *Temporal Coherence in the Perception of Tone Sequences* (Vol. 3). Eindhoven, The Netherlands: Institute for Perceptual Research.

Wagemans, J., Elder, J. H., Kubovy, M., Palmer, S. E., Peterson, M. A., Singh, M., & von der Heydt, R. (2012). A Century of Gestalt Psychology in Visual Perception: I. Perceptual Grouping and Figure—Ground Organization. *Psychological Bulletin, 138*(6), 1172.

Watson, E. (2019). J Dilla's Influence & Legacy Lives on in the World of Lo-Fi Hip-Hop. *Okay Player* [online]. Available at: www.okayplayer.com/music/j-dilla-lofi-hip-hop-influence.html

Weiss, M. (2014). 4 Essential Tips for Mixing Hip-Hop. *The Pro Audio Files* [online]. Available at: https://theproaudiofiles.com/mixing-hip-hop/

———. (2016). Common Uses for Mid Side Processing. *The Pro Audio Files* [online]. Available at: https://theproaudiofiles.com/common-uses-for-mid-side-processing/

Wertheimer, M. (1938). Laws of Organization in Perceptual Forms. *A Source Book of Gestalt Psychology*. Ed. Ellis, W. D. London, England: Kegan Paul, Trench, Trubner & Company, 71–88.

White, P. (2006). Mixing Essentials: The Fundamentals of Mixing. *Sound on Sound* [online]. Available at: www.soundonsound.com

———. (2009). Creating a Sense of Depth in Your Mix: Mix Processing Techniques. *Sound on Sound* [online]. Available at: www.soundonsound.com

———. (2010). Distortion in the Studio. *Sound on Sound* [online]. Available at: www.soundonsound.com

———. (2012). Noise Reduction: Tools and Techniques. *Sound on Sound* [online]. Available at: www.soundonsound.com

Index

Note: Page numbers in *italics* indicate a figure and page numbers in **bold** indicate a table on the corresponding page.

For Product Safety Concerns and Information please contact our EU representative GPSR@taylorandfrancis.com
Taylor & Francis Verlag GmbH, Kaufingerstraße 24, 80331 München, Germany

www.ingramcontent.com/pod-product-compliance
Lightning Source LLC
LaVergne TN
LVHW080655011026
849160LV00014B/1431
* 9 7 8 0 8 1 5 3 5 3 1 9 5 *